普通高等院校对外贸易精品系列教材

# 国际商务谈判实训导学双语版

International Business Negotiation Practice
Learning Guide/Bilingual Edition

（第2版）

主　编　苏　琳

参　编　鄂筱曼　刘　婷　李　伟　罗林敏

　　　　徐钦强　廖皓杰　杨冬玲　谢　涛

　　　　陈　慧　陈　英　谢艳妮

北京理工大学出版社
BEIJING INSTITUTE OF TECHNOLOGY PRESS

## 内 容 简 介

　　本书以国际商务谈判主题为载体，以各个情景练习为主线，将谈判各阶段的工作再现于各章模拟实训情景之中。全书包括谈判主题和四大训练模块，谈判主题包括两个出口谈判主题和两个进口谈判主题，包括卡通立体玩偶出口、棉制服装出口，冷冻厄瓜多尔白虾进口和澳大利亚铁矿石进口。四大模块中，有个人快速反应练习、小组情景陈述练习、小组谈判方案展示练习和小组综合模拟练习。全书实训由浅入深，内容相对独立但又融合贯通，并结合行业背景，对学生的资料搜集能力、方案制作能力、表达能力、应变能力、组织能力、团队合作能力、判断能力和心理素质等进行综合训练和考量。

　　本书既可作为应用型本科国际经济与贸易、国际商务、东南亚经济与贸易、电子商务、商务英语等专业国际商务谈判的实训指导教材，也可作为社会商务人士进行商务谈判训练的自学教材和参考书。

**图书在版编目（CIP）数据**

国际商务谈判实训导学：双语版／苏琳主编． -- 2
版． -- 北京：北京理工大学出版社，2023.3
　ISBN 978 - 7 - 5763 - 2149 - 4

　Ⅰ.①国…　Ⅱ.①苏…　Ⅲ.①国际商务 - 商务谈判 -
双语教学 - 高等学校 - 教材　Ⅳ.①F740.41

　中国国家版本馆 CIP 数据核字（2023）第 036372 号

出版发行／北京理工大学出版社有限责任公司
社　　　址／北京市海淀区中关村南大街 5 号
邮　　　编／100081
电　　　话／（010）68914775（总编室）
　　　　　　（010）82562903（教材售后服务热线）
　　　　　　（010）68944723（其他图书服务热线）
网　　　址／http：//www.bitpress.com.cn
经　　　销／全国各地新华书店
印　　　刷／涿州市新华印刷有限公司
开　　　本／787 毫米×1092 毫米　1/16
印　　　张／10
字　　　数／235 千字
版　　　次／2023 年 3 月第 2 版　2023 年 3 月第 1 次印刷
定　　　价／32.00 元

责任编辑／多海鹏
　　　　　辛丽莉
文案编辑／辛丽莉
责任校对／周瑞红
责任印制／李志强

图书出现印装质量问题，请拨打售后服务热线，本社负责调换

# 前　言

## 一、国际商务谈判实训课程的特点

国际商务谈判是国际商务活动中重要的一环。当前的国际贸易和国际商务活动，存在着种类丰富、供给充分、竞争性强、信息化强的特点。在诸多的同质性或类似的竞争者中，大家的产品或服务不一定具有明显的优势，那么交易的达成更取决于谈判双方在谈判中的表现，如信任感的建立、谈判过程的愉悦程度和沟通的效率等。如果谈判方在这几方面的表现优秀，可以提升自己在对方心中的形象、找到对方的关注点或敏感点、消除对方的顾虑，甚至将"蛋糕"做大，从而大幅提高获得订单的概率。另外，双方在合同履行过程中也可能会遇到各种问题，而问题的解决方式和效率以及今后关系的长久持续，也依赖于双方的谈判能力和技巧。因此，国际商务谈判能力是国际经济与贸易、国际商务、商务英语、电子商务等专业大学生将来进入职场必备的国际商务能力之一。

目前，很多高校相关专业对"国际商务谈判（双语）"课程非常重视，并将授课过程分为理论和实训两个环节。理论课程是对国际商务谈判专业理论知识的学习和掌握，而实训环节则是要巩固学生所学的理论知识，建立起学以致用的思维。要达到这个目的，需要设置特定的情景，赋予学生特定的身份，在相应的行业，用特定的商品参与谈判，才能身临其境、对理论知识有较为深刻的理解、掌握和应用。

国际商务谈判课程的实训过程，需要结合行业背景和专业知识，对学生的资料搜集能力、方案制作能力、组织能力、表达能力、应变能力、合作能力、判断能力和心理素质等方面进行综合训练和考量。因此，国际商务谈判实训课程应具有整体性、实践性和开放性的特点。整体性是指在实训课程的每个阶段，学生能始终处于一个完整的情景，每个谈判实训环节都能实现理论知识、行业背景和个性特征的内在整合；实践性即课程应体现专业技能的需要，教师从知识的传授者、灌输者转变为教学获得的组织者、引导者，学生除了完成老师布置的任务之外，还应从谈判的实践中培养主动发现问题、提出问题、解决问题的能力；开放性是指谈判实训课程应基于学生的需要、动机、兴趣和直接经验来设计，使学生在实训过程中能获得丰富的学习体验，教师的评价标准也具有多元性。

## 二、本书课程的设置和特点

### （一）课程设置

本书的实训程序如图 0 - 1。

1. 个人快速反应练习。在实际的国际商务谈判中，我们在大部分时刻都需要对对方的发言和表现做出快速反应。因此，个人快速反应练习是进行综合模拟谈判前的一项基础练习。要历练的，不仅仅是回答的速度，还有完整的思路、自然的内容衔接以及语言表达的技巧。本环节是以个人为单位进行实训的，共有 25 个情景练习，需要学生在较短的时间内就相应的情景做出积极、有效的反应。

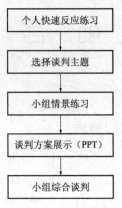

图 0 - 1 实训程序

2. 选择谈判主题。本书共设置中国 PVC 卡通立体玩偶出口、中国棉制服装出口、冷冻厄瓜多尔白虾进口、澳大利亚铁矿石进口四个主题。其中既有学生较为熟悉的商品，也有在中国市场上发展潜力很大，或者对中国的经济发展影响巨大的商品。老师可以根据学生的喜好和每个班级的课时数选择谈判的主题和数量。每个谈判主题中均含有双方身份、产品和市场的基本介绍等与交易有关的主要内容以及相关主题的合同范文。但这些主要内容仅仅是一个轮廓性、指向性的介绍，实训时还需要同学们做详细的资料收集和市场调查工作。

3. 小组情景练习。本环节主要以陈述为主。在国际商务谈判中，有一些陈述的环节，如致欢迎词、致谢词、公司介绍、人员介绍、开局陈述、报价、还价、打破僵局、总结陈词等。这些陈述环节虽然内容不多、篇幅不大，但是技巧性很高。小组情景练习以小组为单位进行练习。通过练习，小组成员的信息收集和筛选能力、文字组织能力、表达技巧以及气场等会得到较大的提升。本环节分为开局、磋商、成交 3 个阶段，包括开局介绍、开局陈述、产品介绍、还价、僵局处理、总结陈词 6 个情景实训项目，全过程都基于一个特定的谈判主题。

4. 谈判方案展示（PPT）。本环节以小组为单位进行。在综合模拟谈判开始前，每个谈判小组都需要进行详细的市场调查，并根据搜集的资料制订一份谈判方案，包括双方优劣势对比、谈判目标、谈判议题安排、人员分工及谈判策略等内容，并以 PPT 的形式展示出来。同时，谈判方案的重点和倾向性等内容和原则，需要通过公司内部会议的形式，由一位或几位会议主持人传递给大家，保证小组所有成员对谈判方案有充分的了解和掌握，这是谈判准备工作的一个重要环节，也是小组成员在综合模拟谈判中有正确或优异表现以及高效分工和配合的基础。谈判方案展示环节，需要每个小组将谈判方案以 PPT 的形式展示。

5. 小组综合模拟谈判。谈判方案明确之后，将进行小组与小组之间的综合模拟谈判。由老师对小组进行配对，并审核双方的谈判目标。谈判成功之后双方应签订一份国际贸易销售合同；谈判失败应说明原因。无论成功与失败，双方都要对谈判的表现、经验和心得进行总结。

6. 其他要求如下。

分组：根据班级人数，建议将一个班分为 4 ~ 6 个小组，其中 2 ~ 3 个小组代表买方，

2~3 个小组代表卖方。在下文中，一个小组意味着代表买方或卖方的一组，一个谈判组即一个买卖双方的组合。

场地布置：谈判教室，有谈判桌、学生计算机、网络，学生自行准备桌牌。

着装要求：正式、简单、大方即可。

（二）本书特点

1. 全面性。在实训中，有个人练习和小组练习。个人要完成快速反应练习，之后小组将以某个谈判主题为依托，完成 6 个情景练习、一个谈判方案制作和展示以及一次完整的模拟谈判。课程呈立体式结构，谈判训练从专项的小模块逐渐转换到综合性的模拟谈判，从个人练习逐渐转换到团队合作。实训的内容和难度由浅入深、由专项至综合，可让学生在逐级加深的训练中全面、立体地理解和掌握谈判的知识。

2. 实战性。在实训中，学生将在现有材料的基础上，查询与谈判有关的各项内容，将其分类、筛选并制作成谈判方案；谈判的内容和目标将以国际贸易合同的条款为基础确定；谈判开始前需要召开内部会议，将双方的优劣势、谈判目标、任务分工、谈判策略等内容进行内部讨论，以便提高模拟谈判的效率和分工合作效果。这些步骤和实际谈判的做法是非常接近的。在如此的铺垫之下，谈判实训将不会限于口舌之争，而是对学生的组织能力、思维、判断能力、口才、应变能力和团队合作能力的综合训练。

3. 双语性。本书有中文、英文两个版本，既适合中文单语教学，也符合双语教学的要求。

## 三、本书的编写

本书由广西财经学院的苏琳主编。广西财经学院的鄂筱曼、刘婷、李伟、罗林敏、廖皓杰、谢涛、陈慧、陈英，广西商务职业技术学院的谢艳妮以及资深商务人士徐钦强和杨冬玲参与编写。同时，本书的编写也得到广西财经学院学生张金显、王鑫、梁晓密、覃佐、黄青妙的大力支持。在此对以上人员表示感谢！

由于时间仓促，书中难免有不足之处，敬请广大读者批评指正。

编　者

# Preface

## 1. Characteristics of International Business Negotiation Training Course

International business negotiation is an important part of international business activities. The current international trade and international business activities have the characteristics of rich varieties, sufficient supply, strong competitiveness and strong informatization. Among many homogeneous or similar competitors, our products or services do not necessarily have obvious advantages, so the conclusion of the transaction depends more on the performance of both parties in the negotiation, such as the establishment of a sense of trust, the pleasure of the negotiation process and the efficiency of communication, etc. If the negotiators perform well in these aspects, they can improve their image in the other party's mind, find the other party's concerns or sensitive points, eliminate the other party's concerns, and even make the "cake" bigger, so as to greatly increase the probability of obtaining orders. In addition, both parties may encounter various problems during the performance of the contract, and the solution and efficiency of the problems and the long-term sustainability of the future relationship also depend on the negotiation ability and skills of both parties. Therefore, international business negotiation ability is one of the necessary international business abilities for college students majoring in international economy and trade, international business, business English and e-commerce to enter the workplace in the future.

At present, many relevant majors in colleges and universities attach great importance to the course of International Business Negotiation (Bilingual), and divide the teaching process into two sections, theory section and practice section. The theory section is to learn and master the theoretical knowledge of international business negotiation, while the practice section is to consolidate the theoretical knowledge learned by students and establish the thinking of applying what they have learned. To achieve this goal, we need to set up specific situations, give students specific identity, and participate in negotiations with specific commodities in corresponding industries, so as to be immersive and have a deeper understanding, mastery and application of the theoretical knowledge.

In the training process of international business negotiation course, students' data collection ability, scheme making ability, organization ability, expression ability, adaptability, cooperation ability, judgment ability and psychological quality need to be comprehensively trained and considered in combination with industry background and professional knowledge. Therefore, the international business negotiation training course should have the characteristics of integrity, practicality and openness. Integrity means that at each stage of the training course, students can always be in a complete scene, and each negotiation practice session can realize the internal integration of theoretical knowledge, industry background and personality characteristics. Practicality means that the curriculum should reflect the needs of professional skills. Teachers roles switch from knowledge imparters and indoctrinators to teaching organizers and guiders. Students should not only complete the tasks assigned by teachers, but also cultivate the ability to actively find, raise and solve problems from the practice of negotiation. Openness means that negotiation training courses should be designed based on students' needs, motivation, interests and direct experience, so that students can obtain rich learning experience in the process of training, and teachers' evaluation standards are also diverse.

## 2. Curriculum and Characteristics of This Book

### 2. 1　Curriculum

The practice procedure of this book is shown in Fig. 0 – 1:

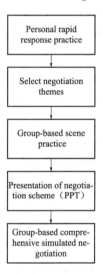

Fig. 0 – 1　Practice Procedure

### 2. 1. 1　Personal Rapid Response Practice

This book has four themes: Chinese dolls export, Chinese garments export, Ecuadorian white shrimps import and Australian iron ore import. These include not only commodities familiar to students, but also commodities with great development potential in the Chinese market or great impact on China's economic development. Teachers can choose the topic and number of negotiations according to students' preferences and the number of class hours in each class. Each negotiation topic contains the main contents related to the transaction, such as the identity of both parties, the basic introduction of products and markets, as well as the contract model of relevant topics. However, the main contents are only an outline and directional introduction. During the practice sessions, students also need to do detailed data collections and market researches.

### 2. 1. 2　Select Negotiation Themes

In actual international business negotiations, we need to respond quickly to each other's speeches and performances most of the time. Therefore, a personal rapid response exercise is a basic exercise before comprehensive simulated negotiation. What we should experience is not only the speed of answer, but also the integrity of ideas, natural contents connection and language expression skills. This session is an individual training, with a total of 25 scene practices. Students are required to make positive and effective responses to the corresponding situations in a short time.

### 2.1.3  Group-based Scene Practice

This session mainly focuses on statements. In international business negotiations, there are some kinds of statements, such as welcome speech, thank you speech, company introduction, team members introduction, opening statement, quotation, counter-offer, breaking the deadlock, closing statement, etc. Although these presentation sessions are short, they are highly skilled. Group-based situational exercises will be practiced in groups. These exercises will greatly improve team members information collection and screening ability, text organization ability, expression skills and aura field. This session includes three stages: opening, bargaining and closing. Specifically, there are six situational practice projects based on a specific negotiation theme.

### 2.1.4  Presentation of Negotiation Scheme (PPT)

This session is conducted in groups. Before the comprehensive simulated negotiation, each negotiation team needs to conduct a detailed market survey and formulate a negotiation plan according to the collected data, including the comparison of advantages and disadvantages of both parties, negotiation objectives, negotiation topic arrangement, work division for team members and negotiation strategies, which will be displayed in the form of PPT. At the same time, the contents and principles of the negotiation plan, such as the focus and tendency, need to be passed on to you by one or more meeting chairpersons in the form of an internal meeting of the company, so as to ensure that all members of the group have a full understanding and mastery of the negotiation plan, which is an important part in the negotiation preparation. It is also the basis for correct or excellent performance, efficient division of work and cooperation of team members in comprehensive simulated negotiation. In the negotiation scheme display phase, each group is required to display the negotiation scheme in the form of PPT.

### 2.1.5  Group-based Comprehensive Simulated Negotiation

After the negotiation scheme is completed, comprehensive simulated negotiation between groups will be carried out. The teacher pairs the group and reviews the negotiation objectives of both parties. After successful negotiation, both parties shall sign an international trade sales contract. If the negotiation fails, the reasons shall be explained. Regardless of success or failure, both parties should summarize the performance, experience and skills gained from the negotiation.

### 2.1.6  Other Requirements

Grouping: According to the number of classes, it is suggested that a class be divided into 4 - 6 groups, of which 2 - 3 groups represent the buyer and 2 - 3 groups represent the seller. In the following, a group means a group representing the buyer or seller, and the negotiation group is a combination of the buyer and the seller.

Venue layout: negotiation classroom, with negotiation table, student computer and network.

Students prepare table cards by themselves.

Dress code: formal, simple and generous.

## 2.2    Characteristics of This Book

### 2.2.1    Comprehensive

During the practice, the individual should complete rapid response practices. The group will complete six scene practices, the production and display of a negotiation scheme and a complete simulated negotiation based on a negotiation theme. The course is in a three-dimensional structure. Negotiation practice is gradually transformed from special small modules to comprehensive simulated negotiations, and from the individual practice to team cooperation. The contents and difficulty degree of the practice are from shallow to deep, from special to comprehensive, so that students can comprehensively and three-dimensionally understand and master the knowledge of negotiation in the step-by-step practice.

### 2.2.2    Practical

In the practice section, students will query each content related to negotiation on the basis of existing materials, classify, screen and make a negotiation scheme. The contents and objectives of the negotiation will be determined on the basis of the terms of the international trade contract. Before the negotiation, an internal meeting shall be held to discuss the advantages and disadvantages, negotiation objectives, work division and negotiation strategies of both parties, etc, so as to improve the efficiency and team members' cooperation effect of simulated negotiations. These steps are very close to the practice of actual negotiations. Under such foreshadowing, the negotiation practice will not be limited to the struggle of words, but a comprehensive practice of students' organizational ability, thinking, judgment ability, eloquence, adaptability and teamwork ability.

### 2.2.3    Bilingual

The book is available in both Chinese and English. It is not only suitable for Chinese monolingual teaching, but also meets the requirements of bilingual teaching.

## 3. Textbook Compilation

This book is compiled by Su Lin from Guangxi University of Finance and Economics. E Xiaoman, Liu Ting, Li Wei, Luo Linmin, Liao Haojie, Xie Tao, Chen Hui and Chen Ying from Guangxi University of Finance and Economics, Xie Yanni from Guangxi International Business Vocational and Technical College, senior business people Xu Qinqiang and Yang Dongling participated in the writing. At the same time, we also received strongly support from students of Guangxi University of Finance and Economics, Zhang Jinxian, Wang Xin, Liang Xiaomi, Qin Zuo and Huang Qingmiao. I would like to express my deep gratitude to above people.

Author

# 目 录

**第一章 模拟谈判主题** ·········································· （1）

谈判主题一 中国 PVC 卡通立体玩偶出口 ·················· （1）

    谈判目标产品 ············································· （1）

    谈判情景设置 ············································· （2）

    案例背景 ·················································· （2）

谈判主题二 中国棉制服装出口 ···························· （6）

    谈判目标产品 ············································· （6）

    谈判情景设置 ············································· （7）

    案例背景 ·················································· （7）

谈判主题三 冷冻厄瓜多尔白虾进口 ······················ （10）

    谈判目标产品 ············································· （10）

    谈判情景设置 ············································· （10）

    案例背景 ·················································· （11）

谈判主题四 澳大利亚铁矿石进口 ························· （12）

    谈判目标产品 ············································· （12）

    谈判情景设置 ············································· （12）

    案例背景 ·················································· （13）

**第二章 个人快速反应练习** ································· （15）

    情景模式 ·················································· （15）

    回答步骤及范例 ··········································· （15）

    情景练习 25 例 ··········································· （17）

    回答思路参考 ············································· （23）

**第三章 小组情景练习** ····································· （28）

    情景模式 ·················································· （28）

发言要求 ……………………………………………………………（28）
情景练习 ……………………………………………………………（30）

## 第四章　模拟谈判方案展示（结合PPT）……………………（35）

实训说明 ……………………………………………………………（35）
主要内容和要求 ……………………………………………………（36）
发言的形式 …………………………………………………………（36）
语言要求 ……………………………………………………………（37）
风险分析 ……………………………………………………………（37）
PPT页面要求 ………………………………………………………（38）
谈判方案展示范例 …………………………………………………（38）

## 第五章　综合模拟谈判 ……………………………………………（42）

准备谈判目标 ………………………………………………………（42）
配对谈判组合 ………………………………………………………（43）
综合模拟谈判 ………………………………………………………（44）
谈判技巧 ……………………………………………………………（44）
谈判中可能会遇到的困境及解决方法 ……………………………（51）
模拟谈判中常见的问题 ……………………………………………（53）

## 附　录 ………………………………………………………………（56）

附录一　国际贸易销售合同范例 …………………………………（56）
附录二　铁矿石购销合同范例 ……………………………………（59）

# Contents

**Chapter Ⅰ　Themes of Simulated Negotiation** ·········································· ( 65 )

Theme 1　Chinese PVC Cartoon Three-dimensional Dolls Export ···················· ( 66 )

Negotiation Target Products ········································································ ( 66 )

Negotiation Scene Settings ········································································ ( 66 )

Background of the Negotiation ···································································· ( 67 )

Theme 2　Chinese Cotton Garments Export ················································· ( 72 )

Negotiation Target Products ········································································ ( 72 )

Negotiation Scene Settings ········································································ ( 73 )

Background of the Negotiation ···································································· ( 73 )

Theme 3　Importing Frozen Ecuadorian White Shrimps ·································· ( 77 )

Negotiation Target Products ········································································ ( 77 )

Negotiation Scene Settings ········································································ ( 78 )

Background of the Negotiation ···································································· ( 78 )

Theme 4　Importing Australian Iron Ore ···················································· ( 80 )

Negotiation Target Products ········································································ ( 80 )

Negotiation Scene Settings ········································································ ( 80 )

Background of the Negotiation ···································································· ( 80 )

**Chapter Ⅱ　Personal Rapid Response Practice** ········································ ( 83 )

Training Mode ························································································· ( 83 )

Answer Steps and Examples ······································································· ( 84 )

Questions of Scene Practice ········································································ ( 85 )

Answer Ideas for Reference ········································································ ( 93 )

**Chapter Ⅲ　Group-based Scene Practice** ··············································· ( 98 )

Profiles ·································································································· ( 98 )

Request for Speech ·················································· ( 99 )

Questions of Scene Practice ·········································· ( 100 )

## Chapter IV  Presentation of Negotiation Scheme（PPT） ··············· ( 107 )

Instruction ························································ ( 107 )

Main Contents and Requirements ···································· ( 108 )

Form of Speech ···················································· ( 108 )

Language Requirements ············································· ( 109 )

Risk Analysis ······················································ ( 109 )

PPT Design ························································ ( 110 )

Examples of Negotiation Schemes ···································· ( 111 )

## Chapter V  Group-based Comprehensive Simulated Negotiation ············ ( 115 )

Prepare Negotiation Targets ········································ ( 115 )

Form the Negotiation Combination ··································· ( 117 )

Comprehensive Simulated Negotiation ······························· ( 117 )

Negotiation Skills ·················································· ( 118 )

Possible Difficulties and Solutions in Negotiation ······················ ( 126 )

Common Problems in the Simulated Negotiation ······················· ( 129 )

## Appendix ··························································· ( 133 )

Appendix 1   International Sales Contract Sample ····················· ( 133 )

Appendix 2   Iron Ore Trade Contract Sample ······················· ( 136 )

# 模拟谈判主题

"国际商务谈判实训"课程需要在特定的情景中完成，因此要选定相关的行业和商品。本章选取了大家比较熟悉的商品和规模比较大的公司，有利于资料收集和角色带入。模拟谈判主题包括两个出口主题和两个进口主题，即中国 PVC 卡通立体玩偶出口谈判、中国棉制服装出口谈判、厄瓜多尔白虾进口谈判、澳大利亚铁矿石进口谈判。第三、第四、第五章的实训环节需要基于本章的某一个谈判主题进行。

本书后面附有相应的国际贸易合同文本，同学们可根据合同文本的内容讨论合作的细节。其中，附录一的《国际贸易销售合同》适合主题一、主题二、主题三，《铁矿石购销合同》是针对主题四的。

提示：阅读每一个谈判主题之后，应思考以下几个问题。

1. 己方和对方的优劣势各是什么？本次谈判对某一方更有利还是供需平衡？

2. 在本次合作中，己方应重点关注和准备的内容是什么？对方可能重点关注和准备的内容是什么？

3. 己方不能让步的和可以调整的内容分别有哪些？对方可能不会让步和可以调整的内容分别有哪些？

4. 谈判中可能遇到的难题有哪些？应将其按难度进行排序。

5. 谈判中什么内容有可能顺利达成一致，应如何安排和利用？

6. 己方还有哪些服务可以提升，以吸引对方的注意，作为成交的法宝之一？例如，打假方案等。

## 谈判主题一　中国 PVC 卡通立体玩偶出口

### 谈判目标产品

谈判目标产品如表 1-1 所示。

表1-1　谈判目标产品-PVC卡通立体玩偶

| PVC 卡通立体玩偶样品 | 产品信息 |
| --- | --- |
| | ● 专业来图定制、来样定制 PVC 卡通立体玩偶<br>● 起订量：5 000 个<br>● 价格：3.2 元/个，FOB 黄浦，价格随数量变动<br>● 产品名称：PVC 卡通立体玩偶<br>● 材质：PVC 等塑胶材质<br>● 产品包装：吸塑、彩盒等（具体可按客户要求专业定制）<br>● 定制品需要提供三维视图或 3D 文件，或寄样定制 |

## 谈判情景设置

### 一、卖方：广东 A 公司

公司于 2012 年成立，现有 3 层厂房，总面积达 8 000 平方米，能顺利完成搪胶、喷油、移印、包装等一系列生产工艺。其中，移印部装配半自动生产设备 50 台，日工作量 80 万次以上；无尘车间，实现食品级生产；自动化包装设备，日工作量 200 立方米以上；滴胶部装配自动化设备，日产量 2 吨以上；资质齐全，公司于 2013 年获得对外出口贸易资质，并通过迪士尼认证、国际玩具业协会（ICTI）认证、环球影视 NBCU 认证、供货商商业道德信息交流（SEDEX）认证、英国零售商协会 BRC 认证和 ISO 9001 认证。

2017 年，公司年产值突破 1 亿元，其海外市场包括中东、东北亚、东南亚、北美、南美和欧洲，主要生产两大类产品：注塑喷油 3D 公仔及 2D 滴胶，滴胶产品主要包括定制印章、定制冰箱贴、定制钥匙扣、定制杯垫、定制开瓶器、定制行李牌、定制 2D 吸盘钥匙扣、定制吊饰、定制笔帽等。公司目前合作的品牌包括华特迪士尼、环球影视、可口可乐、三只松鼠、无损音乐、华为、沃尔玛等。

### 二、买方：法国 D 公司

该公司旗下有几十家小型超市，商品品种齐全。正所谓"麻雀虽小，五脏俱全"。超市销售的商品可以满足居民的基本生活需求；分布较密，多数位于或接近城市居民区；超市内除了销售知名品牌的商品之外，还有 100 多个自主品牌。本次定制商品只用于自主品牌的销售。

### 三、合作背景

双方公司首次合作，这对于中方来说这是一个定制小商品进入法国的好机会；对于法方来说，找到优质的供应商进行长期合作，也是后疫情时代公司可持续发展的重要保障。

### 四、谈判地点：法国

## 案例背景

### 一、后疫情时代全球经济形势和国际贸易特点

后疫情时代，基于疫苗配给、病毒变异、疫情反扑等问题，很多国家的经济复苏出现了不

确定性和分化加剧的特点，全球贸易量也出现萎缩。加上近年来部分国家单边主义和保护主义抬头，以加征关税为特征的全球经济政策不确定性上升，主要经济体之间的贸易摩擦不断。积极的信号是，2020年1月中美第一阶段经贸协议正式签署，为中美两国和世界经济注入了稳定因素，增强了全球市场信心，国际贸易活动逐渐向正常和稳定的方向发展。2021年，全球贸易开始复苏。得益于经济率先复苏，中国在全球产业链、供应链中承担着更大的责任。与此同时，中国也是全球商品的主要"买家"。数据显示，2021年前5个月，中国在进口商品上就花了6.72万亿元，同比增长25.9%。可以说，中国正成为全球经贸的"稳定器"。

资料来源：搜狐网——《海运众生相：堵堵堵，运运运！》，经编者整理改编。

第一，在全球供应链方面，中国具备不可替代的作用。根据中国工程院战略咨询中心2020年年末发布的《2020中国制造强国发展指数报告》，中国制造强国发展指数为110.84，仍处于世界主要制造业国家的第三阵列，质量效益在长时间内仍是我国制造业的较大弱项。但在全球制造业梯队格局中，仍有上升空间。连续11年作为世界最大的制造业国家，中国位于全球生产分工体系中上游的关键位置，拥有全球最齐全的制造产业集群，在对外贸易领域拥有节省成本、有利于创新和深度分工的优势。A公司位于中国广东——中国制造业十强省份的第一名，全省规模以上制造业增加值、企业数量均居全国第一，形成7个产值超万亿元的产业集群，创新水平稳居全国前列。2021年8月9日印发的《广东省制造业高质量发展"十四五"规划》提出了广东将进一步推动制造业高质量发展、重塑产业优势，努力打造世界先进水平的制造业基地、全球重要的制造业创新集聚地等发展目标。

第二，受疫情的影响，很多国家的工厂无法正常开工，因而对中国出口物资的依赖性急剧增强。但是，2021年以来各类原材料纷纷涨价，截至2021年8月，塑料、铜、铝、铁、玻璃、锌合金、不锈钢等原材料分别上涨30%以上，大豆及其加工品上涨幅度超过50%。再加上海运拥堵情况严重，导致"一箱难求"，海运费也节节上升，与原材料价格上涨几乎同步，使国际贸易成本大幅提高。

资料来源：《南方都市报》——《金龙鱼披露半年业绩》，经编者整理改编。

---

**为什么制造产业集群可以节省成本、有利于创新和深度分工呢？**

1. 在产业集群内，生产的上游企业如原材料供应商、设备供应商等数量众多、距离较近，便于生产企业的全面对比和择优选择，从而能够以更低的成本、更快的速度投入生产。

2. 在产业集群内，生产的中下游企业如国内外的用户、经销商也汇集于此。他们为生产企业提供了各种最新的市场信息、产品信息和技术信息。因而，产品集散地同时也是信息集散地，产业集群内的企业可以更快、更全面地了解国内外同类产品的信息并据以快速调整产品结构和改进花色品种，得到创新的主动力和灵感源泉。

3. 在产业集群内，因为有数量众多的上中下游企业和信息资源，每个企业都可以将非重点业务，如其他零部件的生产、产品设计、销售等业务外包，从而将全部资金和精力集中于资金最有优势的那一点，直到将这一点做成全国，甚至全球最便宜的、最好的，形成企业的核心竞争力。

### 著名的苏伊士运河堵船事件

2021 年 3 月 23 日，一艘中国台湾长荣海运股份有限公司的 20 万吨长赐号货轮（Ever Given），凭一己之力堵住了世界海运咽喉之一的苏伊士运河，导致双向交通全面瘫痪，对全球航运干扰影响长达 12 天。据德国保险巨头安联集团估算，受货物交付延长等因素影响，运河堵塞令全球贸易每周损失为 60 亿~100 亿美元。图 1-1 所示为当地唯一的一台小挖掘机拯救大货轮的情景。

**图 1-1　小挖掘机拯救大货轮**

通过苏伊士运河的主要物资包括终端产品和原材料两种。终端产品即衣服鞋帽、电子产品、食品等物资；原材料包括石油、天然气、纸浆、铜等。公开资料显示，通过苏伊士运河运输的石油占全部海运石油的 30%，天然气占全球市场的 8%，运河的堵塞影响了这些产品的供应并导致价格上涨。

2020 年年初至 2021 年 8 月中国宁波－美国洛杉矶海运费的变化，如图 1-2 所示。

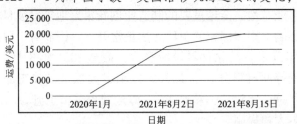

**图 1-2　2020 年年初至 2021 年 8 月海运费的变化**

一个 40 英尺①的集装箱，从中国宁波运到美国洛杉矶，价格从 2020 年 1 月的 1 000 美元左右，涨到 2021 年 8 月 21 日的 20 000 美元左右。疫情期间，全世界的订单像雪花一般飞向中国，但是中国对各国的货物需求数量远远达不到相应的平衡。船公司不愿意空船回程，导致亚洲地区，尤其是中国，形成了"一箱难求"的局面。

根据 CCTV 2"天下财经"栏目发布的数据，2021 年 6—7 月，中国发往欧洲目的港的运价涨幅如表 1-2。

**表 1-2　2021 年 6—7 月中国发往欧洲目的港的运价涨幅**

| 集装箱基准运价 | 2021 年 6 月 25 日 | 2021 年 7 月 1 日 |
| --- | --- | --- |
| 标准集装箱 | 6 600 美元 | 7 300 美元 |
| 大型集装箱 | 12 700 美元 | 13 600 美元 |

资料来源：微信公众号"远方青木"——《不起眼的集装箱，居然成了美国的昂贵物资》，经编者整理改编。

---

①　1 英尺 = 0.304 8 米。

## 二、我国玩具出口特点

中国是世界上最大的玩具制造国和出口国，欧盟是中国玩具产品的主要出口市场。据中华人民共和国海关总署（下称海关总署）统计，2021年1—7月，中国玩具累计出口212.525亿美元，同比增长51.5%；出口目的地以欧洲、美国等传统市场为主，对欧盟出口额为531.4亿元，同比增长75.9%；对美国出口额为851.5亿元，同比增长102.3%。新兴市场包括巴基斯坦、沙特阿拉伯、阿拉伯联合酋长国、哈萨克斯坦、瑞典、南非等，也取得了较快的增长。

随着中国玩具产品在功能、设计及各类IP属性的不断创新发展，未来中国玩具产品在国际市场上的竞争力会越来越强。另外，随着跨境电商等新业态的发展，玩具出口变得更加便利，将给行业整体发展带来新的变化。

## 三、欧洲市场特点和趋势

（一）欧洲市场特点

欧洲有7亿多人口，是中国重要的贸易伙伴，由44个国家组成。英国、德国、法国等西欧国家的人均GDP都超过40 000美元，因而对进口商品的质量特别重视。东欧国家的人均GDP相对较低，更注重产品的性价比。

（二）"中国制造"受到认可

曾经，"中国制造"是廉价与劣质的代名词。但是，经过多年的努力，"中国制造"已成为物美价廉的标志，在国际上拥有独特的地位。疫情之下，人们的工作、出行和生活习惯有了改变，一些"中国制造"的产品，在欧洲成为抢手货。

以自行车为例。为了避免使用公共交通，价格低廉、质量稳定的自行车成为欧洲人的最佳选择。但是，欧洲仅有的一些自行车生产线十分落后，无法满足当地市场的需求。而每年中国制造的自行车约占世界总量的70%。外观精美、质量优良、价格合理的中国自行车，受到欧洲消费者的推崇。据悉，2021年5月中国自行车和电动车对欧出口量暴涨，上万元的"土豪"车型也被抢购一空，生产厂家加班加点，订单仍然排到1个月后。

在医疗产品方面，据德国经济专家分析，自从疫情暴发后，欧洲地区从中国进口的医疗产品，同比增长了50%。在2020年的上半年，欧洲采购口罩的开支就高达140亿欧元，同比增长16.5倍。中国口罩受到欧洲的青睐的原因，除了中国的抗疫工作取得了成绩之外，还有价格的明显优势（中国制造的口罩售价在0.05欧元左右）。中国出售的高级防护设备呼吸机，开始逐渐顶替欧洲本土企业的市场。生产呼吸机的零部件企业，也收到瑞士国内最大的呼吸机制造企业的订单。

（三）疫情期间玩具需求大

疫情期间由于防疫、居家隔离措施的实施，人们居家时间比以前增加了，因而对玩具的需求大增。据美国市场调研机构NPD集团发布的报告，其在12个全球市场追踪的11个玩具类别，2021年上半年销售额同比均有所增长。其中，游戏和拼图增长了59%、户外和运动玩具增长了38%、婴幼儿及学龄前玩具增长了14%。

## 四、RAPEX 玩具产品通报

玩具产品的安全直接关系到儿童的健康和安全，因此世界各国都将玩具列为重点监管产品。近几年欧盟的"安全门"（RAPEX）系统通报中，玩具产品通报的数量都高居"榜首"。

2021 年上半年，RAPEX 共发布 1 116 起通报，玩具产品通报占总通报数的 17 %，对中国的玩具产品的通报共 147 起，占玩具产品总通报数的 78.61 %。但总体来说，中国玩具产品被通报数量同比下降了 32.57 %。

2021 年上半年，共有 17 个国家对中国玩具产品进行了通报。通报量居前五位的分别是：波兰 51 起，占中国被通报玩具产品数的 34.69 %；斯洛伐克 11 起，占比为 7.48 %；法国 10 起，占比为 6.8%；匈牙利 10 起，占比为 6.8 %；瑞典 10 起，占比为 6.8 %。上述 5 个国家通报的中国玩具产品数量共 92 起，占中国被通报玩具产品总数的 62.59 %。

根据 RAPEX 通报风险分类，2021 年上半年中国被通报玩具产品风险类型共计 168 起，有部分产品有多种风险同时被通报。其中，因小部件脱落引起的窒息风险（65 起）和化学风险（65 起）最多。在化学风险中，塑料玩具和包含塑料玩具产品中的邻苯二甲酸酯超标最多。2021 年上半年，我国玩具产品被通报主要原因是不符合欧盟《玩具安全指令》[（EU）2021/1992]、《玩具安全——物理和机械性能测试》（EN71 - 1：2021）、《关于化学品注册、评估、许可和限制法规》[REACH 法规（EC）No 1907/2006]、《有毒金属溶出量测试》（EN71 - 3：2021）、《电动玩具安全标准》（EN IEC 62115：2020）等。

*资料来源：RAPEX 官网 2021 年 7 月 1 日发布的数据，经编者整理改编。*

# 谈判主题二　中国棉制服装出口

**谈判目标产品** ////

目标产品如表 1 - 3 所示。

表 1 - 3　谈判目标产品 - 棉制服装

| 棉制服装样品 | 产品信息 |
|---|---|
| | ● OEM 来样定制棉制服装 10 款<br>● 起订量：1 000 件<br>● 价格：18 元/件，FOB 杭州，价格随数量变动<br>● 产品名称：棉制短袖女性 V 领 T 恤衫<br>● 材质：92% 棉 + 8% 氨纶<br>● 图案：手工钉珠，底部 3D 水浆印花<br>● 颜色：白色、灰色、黑色、玫红色<br>● 尺码：M、L、XL、XXL、XXXL<br>● 产品包装：防水塑料袋（具体可按客户要求专业定制） |

**谈判情景设置**

### 一、卖方：上海 A 纺织品进出口有限公司

上海 A 纺织品进出口有限公司成立于 1959 年，是中华人民共和国成立以后最早设立的国营专业外贸公司之一，也是中国第二家大型综合商社试点单位——东方国际集团的核心骨干企业。公司主要经营高档次、多品种纱、坯布、漂布；腈纶、粘胶纤维织物，呢绒面料、混纺织品和各类服装，纺织制成品以及其他非纺织商品的进出口贸易。自二十世纪八十年代以来，该公司累计出口已逾 80 亿美元，始终位于全国出口企业排行榜的前列。公司贸易范围广，除了各类纺织品及其制成品和其他非纺织商品的进出口业务，承接补偿贸易、来料、来件、进料加工业务，进行技术设备引进和技术交流，接受国内委托及进出口代理业务之外，还涉及国内贸易、服务贸易、房产经营、广告宣传、投资实业、科技开发、咨询服务等诸多领域的业务。

### 二、买方：美国 B 公司

美国 B 公司是服装品牌进口商，在美国各大城市有 36 家分店，经营稳定、其产品具有独特的设计风格，受到美国中青年消费者的欢迎。B 公司计划在中国寻找一家公司长期合作，以 OEM 方式为其生产 10 款棉制服装，对质量有较高的要求，如合作成功，每年将有大量的订单。

### 三、合作背景

双方是在 2021 年 11 月美国迈阿密国际服装纺织品采购展览会上认识的，之后在 B 公司进行了详谈。这是双方的第一次合作。

展会资料来源：中国纺织品进出口商会。

### 四、谈判地点：美国

**案例背景**

### 一、服装出口保持良好增势，但利润空间被挤压

纺织工业是中国传统的支柱产业，在国际上亦有着巨大的竞争力。中国是世界上最大的纺织服装生产大国，也是纺织服装出口大国。受中美贸易战的影响，中国纺织品服装出口经受了一定的压力。2021 年，因为新冠疫情的原因，世界各地的订单回流，使中国的纺织品服装出口出现了新的增势。

海关总署 2021 年 9 月 7 日发布的数据显示，按美元统计，2021 年 1—8 月，纺织服装累计出口 1 984.68 亿美元，同比增长了 5.90%，刷新了同期服装出口纪录。其中服装出口 1 056.95 亿美元，增长了 27.95%。

如果排除口罩因素，上半年纺织品出口实现了正增长，涨幅为 49%。此外，产业链上化学纤维、纱线、织物及纺机等主要商品出口都出现增长。全球经济持续复苏，带动外部需求增加，对中国纺织产品出口起到了提振作用。

但是，这一年来原材料价格大幅上涨、海运成本急剧攀升，国内众多出口企业的经营压力陡增。2021 年以来，纺织服装原料，如棉纱、短纤等价格一路上涨，氨纶价格更是比年

初翻了好几倍，产品仍然供不应求。棉花价格从2021年6—9月累计涨幅超过15%。

## 二、海运物流滞留严重

由于集装箱滞留欧美地区，从2021年6月开始，亚洲出现集装箱"一箱难求"的局面，导致货物运输的周期变长，运费大幅上涨（运费上涨幅度详见主题一）。"现在根本就订不到货柜，要送出去一批货，一般都要一个月到45天。前段时间，我们公司的一批货在港口码头放置了一个多月。"某服装出口企业的谢先生介绍，由于出货周期的拉长，企业的回款周期同样也变长，"以前是20多天就可以收到款，现在至少要40多天。但是，如果服装出口数量比较少，可以选择国际快递和空运专线物流，5~7天到达美国；量大可以选择海运专线，费用十几块/千克，30~40天到美国。"

## 三、我国纺织品服装出口结构

中国纺织品服装出口前五大市场分别为美国、东盟、欧盟、日本、韩国。2021年上半年纺织品服装出口五大市场金额分别为246.4亿美元、224.7亿美元、205.9亿美元、94.7亿美元、44.2亿美元；同比分别增长12.2%、35.7%、−20.0%、−8.4%、18.5%。

美国是中国纺织品服装第一大金额出口市场，2021年上半年有五分之一的产业链终端产品都出口美国。据美国商务部纺织品与服装办公室数据，2021年1—5月美国进口中国纺织品服装金额为101.5亿美元，同比增长了32.3%，继续占领美国第一大进口来源市场。

## 四、来自孟加拉国的竞争

孟加拉国是仅次于中国的世界第二大服装出口国。从2018年开端，孟加拉国接到美国买家的订单量持续增加。美国商务部纺织品与服装办公室发布的官方数据显示，2018年，孟加拉国对美国市场的服装出口收入为54亿美元，而2017年为50.6亿美元。BGMEA前主席Abdus Salam告知《新国家》报，"中美贸易战为孟加拉国的服装出口商创造了一个非常好的时机。孟加拉国服装出口收入在交易抵触中将持续飙升。"Abdus Salam指出，关税导致美国零售商在华收购商品和制造商品的成本升高，为操控成本，不少美国零售商正将收购从我国转向孟加拉国这类成本较节约的制造业中心。

## 五、产业链高端领域的发展仍比较薄弱

目前中国的纺织服装产业的产业链较为完整，综合竞争力较强，但是优势目前仍集中在中低端领域，而绿色环保高端新材料、纺织智能机械的研发等高端领域的发展还比较薄弱。中国早期主要依靠劳动密集型模式来发展纺织品服装产业，但是随着人民生活水平的提升，各种生产要素包括原材料和人力资源等的价格一直在提升，价格优势一直在下降，导致一些订单开始转移到生产要素价格更低的东南亚国家和非洲国家。在复杂的国际形势和国家科技水平提高的背景下，全球纺织产业价值链有重构的趋势。因而，向高端领域迈进是中国纺织品服装产业可持续发展、进一步提升竞争优势而必须完成的艰巨任务。

## 六、美国采购商的六大基本类型

### （一）百货公司买主

很多美国百货公司会自己采购产品，不同品种由不同的采购部门负责。比较大的连锁百

货如梅西百货、杰西潘尼等，几乎在各个生产市场都有自己的采购公司，一般工厂很难打入，他们往往通过大贸易商来选择他们的供货商，自成一个采购系统。该类买主采购量大，价格要求稳定，每年购买的产品变化不会太大，质量要求很高，不太容易变换供货商。这类买主大部分都看美国本地的展览，不会亲自到中国看展。

（二）连锁大型超商卖场

该类卖场如沃尔玛、凯玛特等，采购量大，在生产市场，也有自己的采购系统（Buying Office）。他们的采购对市场价格的敏感度很高，产品变化要求也很大，工厂价压得很低，但量很大。开发力强，价格便宜，资金雄厚的工厂可以进攻这种类型的客户。小厂最好保持距离，否则他们一张订单的周转资金就会让你吃不消，万一质量无法达到验货标准，就难以翻身了。

（三）品牌进口商

品牌进口商大部分是品牌自己进货，如耐克，新秀丽，等等。他们会找有规模，质量好的工厂直接以 OEM 方式下单，他们的利润较好，质量要求有自己的标准，订单稳定，跟工厂建立很长久的合作关系。目前世界上有越来越多的进口商通过中国的展销会来找厂商，这些进口商是值得中小型工厂努力开发的客户。进口商在其国内的生意规模是他们采购数量及付款条件的参考因素。在做生意前，可以通过他们的网站去了解他们的实力。即使是小品牌，也有机会将其培养成大客户。

（四）批发商

一些批发的美国进口商通常采购特定的产品，在国内有自己的发货仓库，通过展览销售他们的产品。量大、价钱低，以及产品的独特性是他们关注的重点。这类客户很容易比价钱，因为他们的竞争对手都是在展览上卖货的，所以价格及产品的差异性要很高。如果是相同的产品，那么常会因为别人价钱较低而跑单。这类进口商主要的采购方式是自己到中国看展采购。例如，很多华人资金较雄厚，他们就在美国做批发生意，成为批发商，回中国采购。

（五）贸易商

这部分的客户，什么产品都可能会买，因为他们有各种不同的客户，所以会采购不同的产品，但订单的延续性比较不稳定。订单量也较不稳定。小厂比较容易做到。

（六）零售商

几年前，美国零售商几乎都在美国本土采购，但商业进入网络化以后，越来越多的零售商自己通过网络询价采购，这类客户其实是很难培养大的，所以比较适合国内批发商来做，工厂只是浪费时间报价，没有潜力。

资料来源：福步外贸论坛——北美市场，经编者整理改编。

---

**中美贸易战的背景及其对中国纺织服装出口贸易的影响**

**中美贸易战开端**

2018 年 1 月，特朗普政府宣布对一些进口的太阳能光伏产品以及大型洗衣机分别进行了最高税率达 30% 和 50% 的关税加征，同时对其他国家经济体进行了关税豁免。美方对华贸易备忘录 "301 调查" 报告显示，美国有关部门将会对约 600 亿美元的中国进口商品进行关税的加征，此举正式将贸易摩擦局势升级为中美贸易战。

> **中美贸易战对中国纺织服装出口贸易的影响**
>
> 自 2019 年 9 月 1 日起，美国开始对当年 5 月份公布的关税清单中的商品进行 25% 的关税加征，而这些商品中涵盖了中国出口美国比重较大的纺织品服装及相关纺织机械。
>
> 美国是中国第一大纺织品服装出口国，此前的纺织服装类关税为 10% ~ 20%，本次加征的关税给中国出口劳动密集型产品带来了巨大的竞争压力，同时加速了订单向越南、印度、巴基斯坦等中低端纺织品服装市场的转移。

# 谈判主题三　冷冻厄瓜多尔白虾进口

## 谈判目标产品

目标产品如表 1-4 所示。

**表 1-4　谈判目标产品 - 冷冻厄瓜多尔白虾**

| 冷冻厄瓜多尔白虾样品 | 产品信息 |
| --- | --- |
|  | 产品 1：<br>厄瓜多尔白虾，40 ~ 50 头/千克 |
|  | 产品 2：<br>厄瓜多尔白虾，50 ~ 60 头/千克 |
| 各进口 200 吨；冷冻集装箱，一般包装：一箱 10 盒，每盒 2 千克 | |

## 谈判情景设置

（一）卖方：厄瓜多尔 E 公司

该公司资信好，与中国多家海鲜进出口企业有贸易往来，每年向中国出口厄瓜多尔白虾 9 000 吨左右。

（二）买方：广东 G 水产有限公司

公司成立时间为 2006 年 3 月 1 日，注册资金为 100 万人民币。公司进口厄瓜多尔白虾多年，已建成完整的水产经营产业链。

（三）合作情况：双方公司第一次合作

（四）谈判地点：厄瓜多尔

**案例背景** ◥◥◥◥

## 一、中国虾类市场

### （一）中国与厄瓜多尔白虾缘分极深

海关总署公布的数据显示，从 2019 年起，中国反超美国，成为全球最大的虾类进口国（2019 年累计进口虾类 72.2 万吨，同比增长 179.8%）。其中，厄瓜多尔、印度、泰国、越南、阿根廷、沙特阿拉伯、加拿大是中国虾类主要进口国。具体来看，2019 年，中国从厄瓜多尔进口虾类达 32.2 万吨，同比增长 321.1%，进口额 18.6 亿美元，同比增长 282.5%。

### （二）中国虾类市场特点

中国有巨大的海鲜消费需求，进口虾占 50% 以上的冷冻虾市场；中国消费者对健康安全的食材诉求越来越强烈，检验检疫部门也加大了检查力度；部分海鲜经营企业已建成包括全球采购、加工、出口、批发的完整产业链；零售市场蓬勃发展，主流电商、商超、加盟店是三个主要的渠道。

## 二、疫情之下厄瓜多尔白虾出口概况

### （一）在中国市场的数量与价格

2020 年前两个月，因疫情的原因，中国进口厄瓜多尔白虾的数量大幅下降。之后在 2021 年 3 月份开始快速复苏，但是在 7—8 月进口数量急剧下降，11 月进口量再次增长，同比增长超过 10%，但是总体呈现波动趋势。2023 年年初，中国调整了进口食品的检测规定，进口虾的数量和价格创历史新高。

在价格方面，2020 年厄瓜多尔白虾的价格下降了 15%～25%，低于印度、印度尼西亚和泰国的白虾。在 2022 年 12 月，厄瓜多尔、越南的白虾价格涨势明显。2023 年年初，厄瓜多尔白虾出口数量持续增长，出口中国的数量同比增长 40% 以上，但是出口价格下跌了 10% 左右。

资料来源：搜狐网——水产网，经编者整理改编。

### （二）在美国市场的动态

2021 年 6 月，美国进口厄瓜多尔白虾的平均价格达到 7.17 美元/千克，比 2020 年 6 月上涨了 19%。迄今为止，厄瓜多尔一直是美国价格最低的白虾供应国之一。

资料来源：水产门户网——微水网，经编者整理改编。

## 三、厄瓜多尔白虾进口通关手续（非疫情情况下）

### （一）准备工作

准备工作包括收货公司备案、中文标签备案，国外生产企业在中华人民共和国国家市场监督管理总局注册。

**（二）出口商应提供**

出口商应提供出口国原产地证明、出口国水产品卫生证书，国外生产企业在中华人民共和国国家市场监督管理总局的注册证书。

**（三）海关监管条件**

对于 HS 编码为 03061730 的商品，海关监管条件为 A/B，即白虾进口需要报检。

A：入境货物通关单；B：出境货物通关单。

**（四）检验检疫监管条件**

对于 HS 编码为 03061730 的商品，检验检疫类别为 P. R/Q. S，报检时需要做出境动植物产品检疫和出口食品卫生监督检验。

P：进境动植物产品检疫；　　Q：出境动植物产品检疫；

R：进口食品卫生监督检验；　　S：出口食品卫生监督检验。

**（五）进口报关申报要素**

进口报关申报要素有品名、制作或保存方法（冻）、状态（带壳、去壳）、拉丁名称、规格（如41～50个/磅）、包装规格。

# 谈判主题四　澳大利亚铁矿石进口

## 谈判目标产品

目标产品如表 1-5 所示。

表 1-5　谈判目标产品-铁矿石

| 铁矿石样品 | 产品信息 |
| --- | --- |
| 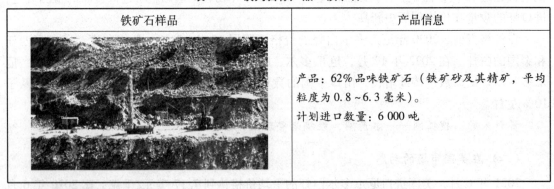 | 产品：62%品味铁矿石（铁矿砂及其精矿，平均粒度为 0.8～6.3 毫米）。<br>计划进口数量：6 000 吨 |

## 谈判情景设置

**一、卖方：H 公司的谈判代表**

详情可参考 H 公司官方介绍。

**二、买方：A 钢集团的谈判代表**

该公司为中国钢铁行业的龙头企业，中国铁矿石的主要购买商、世界铁矿石主要进口商之一。

### 三、合作情况：多次合作

双方公司已多次合作，但双方谈判代表是第一次打交道。

### 四、谈判地点：澳大利亚

**案例背景**

#### 一、中国铁矿石进口量和进口来源

（一）中国是全球最大的铁矿石买家

中国一直都是全球铁矿石最大的买家。2020年，中国新型基础设施建设投资增加提振需求，铁矿石下游行业钢铁行业需求量增加，带动行业整体需求增加。按照兰格钢铁研究中心的测算，2020年中国进口铁矿石11.7亿吨，铁矿石的对外依存度达到了82.3%。且随着国内钢厂产能持续走高，铁矿石的价格也再创新高。

（二）进口集中度高

中国铁矿石进口来源国有30多个，但主要集中在澳大利亚和巴西两个国家。2014年以来，中国进口的铁矿石来自澳大利亚与巴西的占总量的80%左右，且比例一直在升高。2020年，中国从澳大利亚进口的铁矿石占全球进口总量的66%，从巴西进口的铁矿石占全球总量的21%，形成双寡头垄断的局面。

资料来源：证券之星，经编者整理改编。

（三）国家采取措施减少铁矿石进口量

中国推出包括实现钢铁企业所需的原材料多元化、压缩粗钢产量以减少铁矿石使用量、进一步打造海外权益铁矿山等在内的相关措施，以此来减少对澳大利亚铁矿石的进口需求。海关总署的数据显示，2021年1—7月，中国进口铁矿石6.5亿吨，同比减少1.5%。其中，7月份进口铁矿石8 850.6万吨，同比减少21.4%，为连续第2个月下降；环比减少1%，为连续第4个月下降。

中国冶金矿山企业协会秘书长姜圣才指出，根据我国铁矿资源条件、生产条件、技术水平和其他发展条件，未来具备一定的增产能力。只要有政策支持，措施到位，国内矿完全有能力保障2025年、2030年、2035年总需求量的25%、35%、40%。

资料来源：搜狐网——今日钢铁，经编者整理改编。

#### 二、澳大利亚经济对中国的依赖

在出口方面，中国是澳大利亚最大的出口目的国。澳大利亚2017—2018财年的数据显示，澳大利亚对中国的出口占总额的30.6%，主要是矿产等资源和教育服务的需求，总价值为1 946亿澳元。

在进口方面，中国也是澳大利亚最大的进口来源地。2017—2018年，澳大利亚从中国购买了包括电信设备和零部件、计算机、家具、床垫、坐垫、婴儿车、玩具、游戏和体育用

品等价值713亿澳元的商品和服务，占其进口总额的18%。

2020年，普华永道首席经济学家杰里米·索普（Jeremy Thorpe）在"视觉中国"上发布了一篇论文，阐述了如果中国经济出现"硬着陆"，即GDP增速突然下降3%～5%，可能在南太平洋掀起一场"惊涛骇浪"：澳大利亚GDP将直接损失1 400亿澳元（7%），并减少55万个工作岗位。

资料来源：新浪财经，经编者整理改编。

### 三、铁矿石进口通关手续

（一）出口商提供

出口商应提供原产地证明、正本提单、合同、商业发票，及装箱单。

（二）海关监管条件

HS编码为26011120的商品海关监管条件为7A，即进口铁矿石需要办理自动进口许可证和报检。

7：自动进口许可证；A：入境货物通关单。

（三）检验检疫监管条件

HS编码为26011120的产品进口检验条件为M，即铁矿石进口需要检验。

（四）进口报关申报要素

进口报关申报要素有品目、用途、加工方法、外观、成分含量、平均粒度、来源（矿区名称），及签约日期。

# 个人快速反应练习

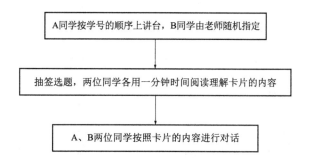

## 情景模式

1. 训练组合：每道题需要 A、B 两位同学一起完成。其中 A 同学为受训者，按学号的顺序完成练习；B 同学仅作配合，可由老师随机指定。

2. 题目：以卡片的形式呈现。卡片需要老师提前制作，每一题有 A、B 两张卡片。A 卡片给 A 同学阅读，B 卡片给 B 同学阅读。其中 A 卡片将说明题目的背景、己方的立场和发言内容，B 卡片为对方的回应内容。

3. 抽签：本章共 25 题，一人一题。每位同学将通过抽签的方式选择自己的题目。A、B 两位同学一起站在讲台上，A 同学通过抽签的方式选择题目之后，两人各自拿好自己的卡片。

4. 阅读卡片：两人各用一分钟左右阅读和理解自己卡片上的内容。

5. 对话：A 同学先说出 A 卡片上的发言内容，然后 B 同学说出 B 卡片的发言内容。对话应流畅、自然。B 同学读完之后，A 同学需要在较短的时间内回答 B 同学。A 同学可以带上纸笔理清思路，思考时间不超过一分钟。

## 回答步骤及范例

### 一、回答步骤

在谈判中，当对方提出一个对我方不利的话题，或者我方比较难接受的要求，要如何回答才能把场面稳住，既能有力地回击、又能让对方听得进去，却不会惹恼对方？本书建议采用如下步骤："认同—但是—表示"。

首先，我们应认同对方的观点。毕竟事物都有两面性，对方只是从他们自己的角度出发来理解这件事情。我们对对方表示理解，等于给对方一个台阶下，让他们从心理上对我们产生共鸣和认同，然后才会接受我们接下来的观点。

其次，表达我们的难处，或者问题所在。在受到认同的基础上，对方听到我们陈述的事实后，才能听得进去，甚至从心理上感到认同、理解。

最后，提出我们的观点，对对方的观点予以反驳。

---

**开心一刻**

温斯顿·丘吉尔是个非常了不起的人，但是有一个很大的毛病——爱喝酒。所以他总是和提倡禁酒的阿斯托夫人斗嘴。

一天，阿斯托夫人走上前来，说道："温斯顿，你又喝醉了，真让人讨厌。"

温斯顿·丘吉尔说："阿斯托夫人，你说得一点也没错，我的确喝醉了。但到了早上，我就会醒过来，而你却一直会让人讨厌下去。"

---

## 二、使用缓冲式语言

我们既希望能给对方一个周全的回答，也需要有一点时间理清思路，那么，可以说一些缓冲式的语言，让对方给我们一些思考的时间。例如，"我明白你的意思，只是不太赞同。给我一点时间，我理理思路。"再如，"我理解你的难处，只是我也很为难。让我想想，我该怎么做。"

## 三、回答范例

● 卡片 A 内容

背景：你是卖方，在谈判会议室。你的公司不算大，但报价已经非常合理，不可再让步。

发言："对于 T1258 号产品，我方最终报价是一套 68 美元，CIF 迪拜。"

● 卡片 B 内容

发言："您的报价还算合理。但是请看看 A 公司的报价。虽然比贵公司高 1%，但人家毕竟是有知名度的大公司，合作起来毕竟放心。让我们再考虑考虑。"

---

**A 的回答**

"（认同）很高兴您收到了 A 公司的报价。看来我们的竞争很多呀！（但是）不过，1% 可不是个小数目，我们做生意的都是要精打细算的，如果提供的服务和质量是一样的，我想您也不希望付出更高的价钱，对吧？再说了，大公司的架子是很大的，如果在订单的执行过程中您需要有什么改动，A 公司是不会同意的。（表示）而我们却很珍惜每一个合作的机会，我们愿意比 A 公司付出更多的精力，斟酌每一个细节，让我们的每一个客户都享受较好的待遇。虽然是小公司，但是我们也在业界经营多年了，不会为了一笔订单损害我们的名誉。您完全可以放心。"

以下是上述案例的回答参考思路。

1. 大公司也是从小公司发展起来的。

2. 小公司也会在意名誉。

3. 国家正在扶持中小企业的发展，对很多国家来说，中小企业对 GDP 的贡献率是很高的，如美国、意大利。

4. 小公司很珍惜每一次合作的机会，对今后的合作可以做灵活调整。

5. 1%，对整个订单而言，也不是一笔小数目。

6. 大品牌也一样会有丑闻。

7. 我们的原料来源和 A 公司是一样的/我们的……和 A 公司是一样的。

注意：反驳对方可以有很多个思路，但是回答中，有 1～2 个重点思路即可，切勿将所有思路堆砌。

## 情景练习 25 例

### 第一题　接机

• 卡片 A 内容

**背景：** 下午三点，在接机现场。你是谈判的负责人之一。希望你能给对方一个好印象，最好能联系到双方的合作或公司的业务。

**发言：** "很高兴见到您，艾伯特先生！飞机是否舒适，旅途愉快吧？"

• 卡片 B 内容

**发言：** "谢谢你过来接机，旅途不错。我第一次来中国，中国的发展比我想象得好多了！"

### 第二题　布展

• 卡片 A 内容

**背景：** 你是卖方，在展览会现场。你公司在这个行业发展只有 3 年，时间不长。但公司有经验丰富的专业人士和勤勉的员工，请把握这个机会，让对方有去你公司详谈的意愿。

**发言：** "看得出史密斯先生对我们的产品很有兴趣。明天到公司详谈如何？"

• 卡片 B 内容

**发言：** "产品确实不错，但是你们的布展过于简单，看来经验不足啊。贵公司在行业内应该是新手吧？"

### 第三题　心理战术

• 卡片 A 内容

**背景：** 你是卖方，在谈判会议室。你公司是大公司，实力雄厚；对方公司是小公司。虽然你的报价不是市场最低，但也在合理范围内。成交价格关乎你的业绩和在公司里的地位，不能再让步。

**发言：** "经过专业的价格核算，我们的最终报价是每件 25 美元，CIF 长滩。"

• 卡片 B 内容

**发言：** "这个价格不算很高，但您也知道，小公司赚不了多少钱。像贵公司这样的大牌企业，有成本和渠道的优势，服务才是您重点考虑的因素吧？再降2%对您来说应该不算什么，但必能让我们全力以赴了。"

### 第四题 运输方式

● 卡片A内容

**背景**：你是买方，在谈判会议室。目前是海运高峰期，你们希望分批装运，允许转运。回答时请坚持立场。

**发言**："不知在运输方式上，贵方有什么建议？"

● 卡片B内容

**发言**："我方的要求是，允许分批装运，不可转运。"

### 第五题 犯错误

● 卡片A内容

**背景**：你是卖方，在谈判会议室。你公司的资历、实力不错。谈判初期，谈判正在有条不紊地进行。但此时对方一直拿着你方提供的资料，窃窃私语，不置可否。请在承认错误的前提下巧妙圆场。

**发言（微笑）**："不知贵方在讨论什么，能否告知一二？"

● 卡片B内容

**发言**："不是我挑剔。你看，这个明细单打错了一个字，我们不得不对你们的执行能力表示质疑。"

### 第六题 价格分解

● 卡片A内容

**背景**：你是卖方，在谈判会议室。产品是西服；订单A是新款，订单B是去年的旧款，订单数量各为200件，两款西服的质量和面料略有差异。对方发言后请坚持立场。

**发言**："这是订单A和订单B的报价详情，请您详阅。"

● 卡片B内容

**发言**："订单A比订单B贵了2 000美金！你不是开玩笑吧？通货膨胀也没有这么高呀！"

### 第七题 货损

● 卡片A内容

**背景**：你是卖方，在谈判会议室。合同中注明要冷冻集装箱运输，你方租赁的确实是冷冻集装箱，但是由于船方的问题导致损失。请解释，让对方息怒并一起讨论解决方案。

**发言**："货物发生了问题我们也感到很震惊。具体损失是多少？是如何发生的？"

● 卡片B内容

**发言（生气）**："合同中注明要冷冻集装箱运输。可是30%的货物因温度过高而变质了。请给个解释。"

### 第八题　压价

● 卡片 A 内容

**背景：**你是买方，在谈判会议室。要求再降1%，被拒绝之后要坚持立场，但不要得罪对方。

**发言：**"这个价格还是太高了，再降1%，马上签合同！"

● 卡片 B 内容

**发言：**"这个交易条件对我们来说真是史无前例了。若不是我国经济不景气，我们无论如何都不会降到这个程度。就是赚点喝粥的钱，再降就要喝西北风了。"

### 第九题　吹毛求疵

● 卡片 A 内容

**背景：**你是买方，在谈判会议室。请用吹毛求疵策略对该产品进行挑剔以便降低价格；这款洗衣机的品牌在中国还没有多少知名度。

**发言：**"这款洗衣机有什么性能，请具体介绍一下。"

● 卡片 B 内容

**发言：**"这款洗衣机可以对水进行加温，提高洗衣服时的去污能力；通过滚动滚筒来达到洗涤效果，无须依赖大量的水流进行洗涤，能节约用水；带有烘干功能，对中国南方的潮湿天气来说是一项非常好的功能。"

### 第十题　准备不充分

● 卡片 A 内容

**背景：**在谈判会议室，你是冰箱销售方。你知道韩国品牌在美国家电市场很有影响力。但具体的情况你们没有调查过，不知道比例是否如此之高。

**发言：**"这个价格已是我方底线，不知贵方还在顾虑什么？"

● 卡片 B 内容

**发言：**"在美国家电市场，韩国品牌的影响力日益扩大。2015年，4 000美元以上法式落地双扇玻璃门冰箱的77%都是三星电子的产品。要推广中国品牌的冰箱，我们需要花费更多的成本和时间。这个价格，我们实在是划不来。"

### 第十一题　发脾气

● 卡片 A 内容

**背景：**在谈判会议室。你的上级是位女性。公司高层正在对此事进行紧张的讨论。对于对方挑衅的语言，你可以自己权衡如何处理，可以生气、可以耐心。

**发言：**"不好意思，上级还在讨论这个决议，请您再耐心等待一会。"

● 卡片 B 内容

**发言（生气）：**"你们的上级是个女人吗？磨磨蹭蹭的，一个请示一天都没有得到答复。我们大老远过来，可不是专门来吃北京烤鸭的！"

### 第十二题　付款方式

● 卡片 A 内容

**背景：** 你是买方，在谈判会议室。你公司资历较好，希望能以30% T/T方式支付。请尽量说服对方。

**发言：** "我公司一向都是30% T/T付款的。上个星期签了一个大单，比你们的数量还大，也是T/T付款的方式。"

- 卡片B内容

**发言：** "不好意思，我们公司只接受不可撤销即期信用证支付方式。"

### 第十三题　意外的话题

- 卡片A内容

**背景：** 你是买方，在谈判会议室。对方看你方的产品成本核算清单，思考良久并提出尖锐问题。请坚持立场，予以反驳。

**发言：** "布雷迪先生沉默良久，不知对这份清单有何看法？"

- 卡片B内容

**发言：** "成本清单没什么问题，但重点不是这个。我们到中国超市看过，产品没有放在超市的主过道上，而是在一个小角落里。这么说吧，产品利润微薄，是你们营销工作做得不到位，和我们没关系。价格降得再低也没有用。"

### 第十四题　共同的责任

- 卡片A内容

**背景：** 你是红酒购买商，在谈判会议室，讨论××市场打假的问题。双方计划使用意大利的ID软木塞，你希望软木塞的购买由对方负责，且使用软木塞增加的成本由双方共同承担。但是对方不同意承担成本。请尽力说服对方。

**发言：** "软木塞必须在灌装时使用。这样吧，软木塞的采购就由你们负责。意大利ID软木塞的成本比一般软木塞要高一些，增加的成本由双方共同承担，如何？"

- 卡片B内容

**发言：** "就××市场的假酒最多。其他的客户对ID软木塞并没有需求，共同采购没有问题，但是增加的成本我们可不承担。"

### 第十五题　发货时间

- 卡片A内容

**背景：** 你是卖方，在谈判会议室。对方的订单，最快下周三能发货。如果对方提出的发货时间早于下周三，请拒绝，但要尽力留住客户。

**发言：** "现在是杏仁的销售旺季，订单量很大。但我们能保证下周发货。"

- 卡片B内容

**发言：** "圣诞节快到了，我们的客户急需这批杏仁。如果下周一不能发货，我们的损失会很大。"

### 第十六题　额外要求

- 卡片A内容

**背景**：你是卖方，在谈判会议室。成交在望，但是对方突然提出延长一年产品保质期的要求，你方不可接受。回答时需要注意言辞，不要导致谈判重点重启，成交遥遥无期。

**发言**："价格已经很实惠了，实在找不到拒绝的理由啦！"

● 卡片 B 内容

**发言**："这样吧，大家都爽快一点！你们的价格我们接受，你们延长一年产品的保质期就好。"

### 第十七题 接受

● 卡片 A 内容

**背景**：你是买方，在谈判会议室。你们的最终目标是让对方再降价 1.2%。若对方答应，请接受，但不要太直接。

**发言**："再降价 1.5%，成交！"

● 卡片 B 内容

**发言**："这个星期谈得很辛苦，我们都看到了彼此的诚意和努力。这样吧，为了愉快的合作和长久的发展，我方最后降价 1.2%，你们再不同意，我们也只好打道回府了。"

### 第十八题 离间计

● 卡片 A 内容

**背景**：客场谈判，现已进入关键时刻，你是总经理，晚上对方主谈请你出来喝酒。对方可能使出离间计，请不要上当，并巧妙回应。

**发言**："谢谢你请我出来喝酒。好久没那么放松了，这段时间可真累呀。你也累坏了吧，来，喝一杯。"

● 卡片 B 内容

**发言**："这里确实很放松啊。你也是个坦率的人，所以有句话我不得不说。你们主谈陈先生的想法和你不一样，他把短期利益看得太重，才导致了今天的局面。陈先生是个人才，但是他的立场和态度，让我对本次谈判不太看好。"

### 第十九题 合作程度

● 卡片 A 内容

**背景**：你是冰箱销售方，在谈判会议室，双方在讨论冰箱的消费者体验问题。你希望对方参与活动的筹划和举行，并承担 10% 的费用。对方愿意配合但不愿意承担费用。请坚持立场并强调投入与不投入在工作责任感和效果等方面的差异。

**发言**："消费者体验活动是我们下一阶段产品研发的基础，也是贵公司进一步稳固和开拓市场的基础。在活动的筹划和举办方面，我们希望贵公司能与我们合作，并承担 10% 的费用。"

● 卡片 B 内容

**发言**："消费者体验活动确实重要，但你们才是受益者。我们可以配合，但不会承担费用。"

### 第二十题　中间人

- 卡片 A 内容

**背景：**你是卖方，在谈判会议室，客场谈判。陷入僵局，对方总经理出差了。你和对方律师沟通，动之以情，晓之以理，希望他能帮忙调和僵持的状态。

**发言：**"我认为总经理先生的离开是个策略，是故意的。我们可以让步，但要成交，必须双方一起让步。"

- 卡片 B 内容

**发言：**"我能理解你方的立场，我也不想看到本次谈判徒劳无功。但是总经理有他的打算，我也不好干预。"

### 第二十一题　铁矿石

- 卡片 A 内容

**背景：**你公司是铁矿石买方，中国钢铁行业的龙头企业。在谈判会议室。目前，中国钢材市场亏损严重，为了避免中小钢铁企业大规模倒闭，你公司不得不两次提高钢材价格。在本次谈判中，铁矿石价格没有降到理想的程度，必须坚持立场。

**发言：**"史密斯先生应该也清楚，互利共赢才是长久发展的基石。现在中国钢材市场亏损严重，铁矿石原材料价格这么高，有谁会买呢？"

- 卡片 B 内容

**发言：**"亏不亏损我可不知道，我只看到中国钢材价格今年已经两次提价，原材料凭什么降价？"

### 第二十二题　场外谈判

- 卡片 A 内容

**背景：**你是买方，谈判进入尾声。你觉得价格还是太高。在签约前的晚宴上，你与总经理聊得很合拍，但谈判是与总经理助手进行的。你感觉价格偏高。总经理为人豪爽，你感觉还有回旋的余地。请把握机会再降一点价格。

**发言：**"我讲了许多道理欲让您的助手理解，没想到他的理由比我还多，反过来让我去理解他。这个价格我们就算接受，也是心不甘情不愿。"

- 卡片 B 内容

**发言（总经理）：**"哈哈，我这个助手确实比较强势。不过，彼此理解不正是成交的基础吗？"

### 第二十三题　最后通牒

- 卡片 A 内容

**背景：**你是买方，在谈判会议室。主场谈判接近尾声。价格接近目标水平，对方表示要离开。请坚持你的目标，也要给对方台阶下。

**发言：**"谈了这么久，我知道贵方的诚意。但价格实在偏高。"

- 卡片 B 内容

**发言：**"唉，我做了巨大的让步，谈判也没有成功。再拖也没有意义了，我坐明天的飞机回国，有什么情况再说吧。"

### 第二十四题　加快进程

• 卡片 A 内容

**背景：** 你是卖方，在谈判会议室。你发言之后通过对方表情的变化得知，对方已接受你的条件。但他们不会直接表态。你需要促进交易的达成。

**发言：** "为了长远合作，我建议技术费为 T 国市场价格的 85%。"

• 卡片 B 内容

**发言：** "贵方的确做了很大让步，我也感到贵方的合作诚意。不过，我方需研究之后，才能答复贵方。"

### 第二十五题　拒绝降价

• 卡片 A 内容

**背景：** 你是卖方，在谈判会议室。谈判接近尾声，价格已经非常合理。但对方要求再降 2%，这是很不合理的。请反驳，并坚持立场。

**发言：** "本次谈判很愉快，我们彼此信任和欣赏。您知道我们是没法降 2% 的，何必节外生枝呢？"

• 卡片 B 内容

**发言：** "美国市场的消费能力和潜力，我相信贵公司是了解的。这不是价格的问题，是双方长期合作的问题。中国有个词语叫作'让利'做生意，讲诚意。我希望能看到你们的诚意。"

## 回答思路参考

### 第一题　接机

1. 肯定中国的发展。

2. 联系到行业或公司的发展。

3. 相信合作成功。

### 第二题　布展

1. 承认公司入行不久。

2. 强调公司在产品的品质和服务质量上是有保证的。

3. 布展产品只是公司成果的一部分。

4. 想要更全面地了解产品情况，请到公司详谈。

### 第三题　心理战术

1. 表示理解。

2. 大公司在经营方面也是精打细算的。

3. 如果不降价 2%，你方就不会全力以赴吗？

### 第四题　运输方式

1. 理解对方的要求。

2. 但是，在目前的情况下，这样的要求是很难实现的。

3. 我们做了许多，都是为了……

4. 结尾（结尾很重要，它关系到对方回答的内容、态度和立场，因此应该是积极和合作的）。

### 第五题　犯错误

1. 承认错误，表示道歉，表示整改的态度。

2. 希望对方不要以一个错误来质疑我方的整体执行能力。

3. 今后……

4. 结尾（结尾很重要，它关系到对方回答的内容、态度和立场，因此应该是积极和合作的）。

### 第六题　价格分解

1. 理解对方。

2. 一般新款都比旧款贵。

3. 分解价格。200 件总共提价是 2 000 美元，一件只贵了 10 美元。

4. 新款是市场的刚性需求。

5. 结尾（结尾很重要，它关系到对方回答的内容、态度和立场，因此应该是积极和合作的）。

### 第七题　货损

1. 请息怒。

2. 该承担的责任，我们绝不逃避。

3. 我方确实租赁了……

4. 结尾（希望能好好讨论，降低损失）。

### 第八题　压价

1. 理解对方的困难。

2. 不成交才是真正的喝西北风。

3. 我们都是怀着真诚的态度过来谈判的……

### 第九题　吹毛求疵

1. 听起来不错。但是……

2. 加温……

3. 烘干功能……

4. 结尾（结尾很重要，它关系到对方回答的内容、态度和立场，因此应该是积极和合作的）。

**第十题　准备不充分**

1. 肯定韩国品牌的冰箱确实受欢迎。

2. 但是市场不是……

3. 你方真正需要的是……

4. 结尾（结尾很重要，它关系到对方回答的内容、态度和立场，因此应该是积极和合作的）。

**第十一题　发脾气**

1. 对拖延表示道歉。

2. 解释原因。

3. 对性别歧视表示抗议。

4. 今晚请对方吃北京烤鸭。

**第十二题　付款方式**

1. 肯定信用证的好处。

2. 我方的资历……

3. 信用证的局限性：单据不能出现错误；对买方的资金占用量大等。

4. 结尾（结尾很重要，它关系到对方回答的内容、态度和立场，因此应该是积极和合作的）。

**第十三题　意外的话题**

1. 赞扬对方的工作很详尽。

2. 营销工作会跟进和改善。

3. 货架的摆放并不是产品推广的关键因素。

4. ……才是本次合作的关键。

5. 结尾（结尾很重要，它关系到对方回答的内容、态度和立场，因此应该是积极和合作的）。

**第十四题　共同的责任**

1. 承认××市场假酒多。

2. 但是××市场也是最大的。

3. 我们只有……才能利益最大化。

**第十五题　发货时间**

1. 我们也希望能尽量满足你们的要求。

2. 现实的困难……

3. 下周三一定能发货。

4. 结尾（结尾很重要，它关系到对方回答的内容、态度和立场，因此应该是积极和合作的）。

### 第十六题　额外要求

1. 直接表示失望。

2. 双方都付出许多……为何在成交时节外生枝？

3. 结尾（结尾很重要，它关系到对方回答的内容、态度和立场，因此应该是积极和合作的）。

### 第十七题　接受

1. 沉默一会。

2. 1.2%……真的很难交差。

3. 但是，谈了这么久，我们也能感觉到你方的诚意……

4. 接受……希望在今后的合作中……

### 第十八题　离间计

1. 对出现的情况表示意外。

2. 请对方不要担心。陈先生的观点和团队是一致的。

3. 也会重视对方的想法，会回去和陈先生再次统一立场。

4. 结尾（结尾很重要，它关系到对方回答的内容、态度和立场，因此应该是积极和合作的）。

### 第十九题　合作程度

1. 明白对方的想法。

2. 双方都是受益者。

3. 涉及利益的活动，才会让人全心全意地参与。

4. 利益最大化。

### 第二十题　中间人

1. 理解对方的立场。

2. 强调他的作用。

3. 谈判双方，包括总经理，都是希望谈判成功的。

### 第二十一题　铁矿石

1. 承认钢材价格提价。

2. 提价了也是亏损。

3. 采购铁矿石是为了支撑将来的市场，而不是如今的市场。

4. 对方是专业的。

5. 结尾（结尾很重要，它关系到对方回答的内容、态度和立场，因此应该是积极和合作的）。

**第二十二题　场外谈判**

1. 彼此理解这个观点不错。

2. 但是更希望得到总经理本人的意见。

3. 价格还是偏高了。希望总经理能做主。

4. 愉快地成交才是最好的成交。

**第二十三题　最后通牒**

1. 想不到谈判会进行到这一步。

2. 还是希望对方能再坚持一会，毕竟……

3. 安排一下明天的活动，观光、酒会、聚餐等，让对方改期回国。

**第二十四题　加快进程**

1. 好！我们相信很快会有好消息！

2. 我们也趁这个时间，再斟酌合同的其他条款。

3. 结尾（结尾很重要，它关系到对方回答的内容、态度和立场，因此应该是积极和合作的）。

**第二十五题　拒绝降价**

1. 太为难我了。

2. 我们是否有诚意，您还看不出来吗？

3. 我们很重视美国市场。

4. 望珍惜……

# 小组情景练习

本环节以小组为单位，依据谈判的开局、磋商和成交三个阶段，设置6个情景。在每个情景中，每个小组根据题目的要求搜索相关资料、确定发言的内容，并做出相应的情景演示。

本章需以第一章中的某一个谈判主题作为练习的基础。

## 情景模式

1. 分组：全班分成6~8个小组（每个小组5~7人较为合适，小组的数量视班级人数而定），其中3~4个组担任买方的角色，3~4个组担任卖方的角色。每个情景都以小组为单位进行演示。

2. 小组内部分工：一个小组内部根据情景的需要划分为买卖双方。在演示时，仅一位同学发言，其他同学做配合和营造相应的气氛，不发言。在实际的谈判中，这些情景也是以某个人的发言为主，而非讨论的环节。因此，在课堂练习中，也是以某一位同学的发言为主，其他同学可以在动作、表情上予以配合，营造一定的气氛。

在不同的情景中，每个小组发言的同学应轮换。

3. 小组角色：在本环节的6个情景中，情景3"产品介绍"比较特殊，全班6个小组都以卖方的身份完成练习；另外的5个情景，按照原定的买方或卖方的身份进行。

## 发言要求

> **提 示**
>
> 情景练习的整体效果由"发言的内容"和"发言人的表现"共同构成。其中发言的内容应考虑整体性、重点性、针对性等问题；而发言人的表现，则与仪态、气场、语调、音量、语速、连贯性等因素相关。

### 一、发言的内容

1. 发言的内容应由整个小组共同讨论决定，而不是由发言的同学独自完成的。

2. 发言的内容应与主题直接相关。在搜索的过程中会遇到大量繁杂的资料，其中有部分内容看起来很有诱惑力，但其实与本次谈判没有直接关系，也不是对方关心的内容。因此，需要对搜索到的资料进行筛选，以确保直奔主题、切中要点。

以铁矿石谈判的开局介绍为例，代表 B 集团的小组在搜索集团资料时，会看到很多有关 B 集团钢铁生产规模、钢材产品技术改造等内容，并且这些也是 B 集团引以为豪的内容。有的小组会以此为重点，然而这样的安排并不是谈判对方，即铁矿石的销售方最关心的问题。虽然己方发言的时候颇有些气势磅礴的感觉，但是对方却兴趣索然。为什么呢？首先，谈判的目标产品是铁矿石，而不是钢材；其次，B 集团在国内市场的优势与国际市场的地位和话语权是不能直接画等号的。联系是有的，但是需要我们在表达方面进行一些转换。那么，重点应该是什么，要说什么才能抓住对方的心、引起对方的兴趣，请同学们自己思考和归纳。

3. 发言的内容应考虑双方的需求，包括客观需求和主观需求，如双方的处境、实力对比、对方的顾虑、我方计划营造的气氛、建立需求、引起对方交易兴趣等因素。如在营造气氛方面，应思考我们在这个场景中需要表现的是什么，想给对方压力还是诱惑，想创造什么类型的气氛，轻松还是严肃、坦率还是神秘等。

### 二、语言要求

1. 脱稿。不能以"念"的形式发言，应有一些"说"的语言。例如，"此时，我想我们每一个人都应该用最热烈的掌声来……"再如，"有关今天会议的主要内容，我主要讲述如下四点……"

2. 直接发表演说，不要讲思路。如开局介绍时说："大家好，我是 × 号，×××（名字），下面请允许我介绍 ×××"。

3. 应有一定的深度、广度和专业性。可通过对比以下几句话，体会有深度的语言和非专业的语言的差异。"大规模扩产，稳固了我们在市场中的主导地位。""有关此类问题诉讼案件的增加不仅导致了法律成本的增加，而且分散了我们集团管理层的精力。""我们的质量真的是非常好的。"

4. 有一定的表现力。相同的内容由不同的同学演示，得到的效果可能会差别很大，这与表现力有关。表现力是仪态、气场、语调、音量、语速、连贯性等方面的综合表现。学生的表现力与个人的历练、演讲能力、对谈判的理解有关，希望在这门课程的学习过程中每位同学的表现力都能得到充分的提高。

### 三、准备工作

学生对题目的准备应在课堂上进行，不要因为课时少（有的专业仅有 24 个课时）而让学生在课前准备。若课时不足，可以适当减少题量。

在课堂上准备时，学生可以对比各组的表现，并结合老师对前一题的评价和建议，调整和修正下一题查询资料的方向、内容的筛选和语言的应用等问题，使回答的思路、专业性、表现力等方面得到逐步提高。如在课前准备，则进步的空间不大，且容易造成组内简单分工，即每个人负责一题，得出的结果没有经过小组讨论。受限于个人的思维和水平，一般很难考虑周全、经不起推敲。

### 情景练习

每个情景的发言时间为 3~5 分钟，准备时间为 20~30 分钟。

## 一、开局阶段

商务谈判开局阶段，一般是指双方在讨论具体、实质性交易内容之前彼此熟悉和就本次谈判的内容分别发表陈述的阶段。开局阶段应为谈判创造一个适宜的氛围，可以是轻松的、融洽的，也可以是高压的、严肃的，具体的选择要由谈判的主题而定。成功的开局，可以为后面的谈判奠定一个良好的基础。开局阶段的实训包括开局介绍、开局陈述和产品介绍三个题目，每个小组根据自己的身份（买方或卖方），结合第二章的背景介绍和网络上搜索的资料，确定思路、内容和相应的用语，同时可以增加一些调节气氛的语言。本阶段的实训包含开局介绍，开局陈述和产品介绍三个情景。

**情景 1：开局介绍**

请以主场谈判主持人的身份，向对方介绍公司和谈判的人员，具体内容如下。

1）公司：实力、产品、经营规模等。

2）人员：姓名、职位，谈判中负责的领域、特点（学历、经验、职称等）。

背景：谈判双方的人员是第一次打交道，但之前双方公司已合作六年，因此大家对双方公司的实力、合作情况应该是比较熟悉的，在介绍时应注意内容的选择。

注意事项：

1）作为谈判的主持人，需要在开头和结尾说一些开场白和结束语。

2）在人员介绍中，介绍 3 名成员即可，可使用学生真名，但对应的职位、特点应由学生自拟。

3）发言应紧密结合谈判内容，不要偏题。即我们所说的每一句话，都应有的放矢，也应该是对方关心的，或者是与对方有关的。如有的小组在介绍人员分工时，提到"×××同学负责记录谈判内容"，这是本组内部的分工，就没有必要告诉对方，对谈判的结果也没有什么推动意义。

4）发言内容应精简，注意整体的流畅性。开局介绍的重点是营造对我方有利的气氛，表达诚意、突出我方的优势，给对方一定的压力或诱惑，为将来的成交打好基础。对于公司和产品的信息，不建议将一整段长长的公司资料一五一十地念下来，或者进行详细的产品介绍。这样做，不仅重点优势没有突出，还会导致现场气氛沉闷、听者走神，达不到预期效果。

5）发言应注意语言的应用和契合度。请对比以下几句话。"下面我介绍一下本公司实力。""……从而证明我公司实力。""我公司实力如下……""下面请允许我对本公司进行简单的介绍。"哪一句话听起来更加自然、生动呢？

6）发言应体现己方优势，如买方在供过于求的情况下还强调产量大幅升高，就不合适。

---

**冰箱购买方开局介绍范例**

各位 H 公司的谈判代表，你们好！我是 D 公司销售代表×××，我代表 D 公司向各位谈判代表的到来致以诚挚的欢迎。接下来，请允许我介绍我公司的谈判代表：销售总经理×××、销售代表×××、财务总监×××。

今年已是我们双方公司合作的第六年了，此前的合作都获得了令人满意的结果。很高兴今天我们再次相聚于此。我方本次的谈判人员是公司各部门遴选的优秀人才，希望我们能承前启后，再次和贵公司合作成功，共创佳绩。

贵公司是冰箱制造行业的一流企业，在业界具有良好的声誉。贵公司的新式冰箱，其功能、外观及创新技术的使用，都正好顺应了现在市场的居民需求，销售发展空间潜力大。而以我公司的实力和经验，若能得到贵公司这两款新式冰箱在法国唯一销售代理权，定能将产品成功推向市场。

接下来的几天，希望我们双方能够积极讨论，并有所收获。相信我们双方的合作不仅为贵公司将来在其他地区的销售打下一个稳固的基础，也帮助贵公司在法国市场建立更高的品牌形象。

我方聘请了优秀的中文翻译人员×××，若贵公司在法国有任何需求，都可联系我们，我们将竭尽全力提供帮助。再次欢迎各位谈判代表的到来，并预祝我们在接下来的谈判中取得成功！

资料来源：广西财经学院国际商务 1621 班第三小组，身份：冰箱购买方

---

**情景 2：开局陈述**

开局陈述应对谈判程序和相关问题达成共识。谈判双方分别表明己方的意愿和交易条件，通过对方陈述摸清对方的情况和态度，为实质性磋商阶段打下基础。

内容：应包含立场、关键问题、困难、希望等内容。

注意事项：开局陈述还处于摸底阶段，只需要指出一个方向，不要涉及过于具体的交易内容。

开局陈述范例如下。

背景：印度正朝着超级经济体的目标快速发展，也成为许多全球性计算机和软件公司建立工厂和研发机构的选择。邦加罗尔，印度南部的城市，吸引了国际一流的信息和科技公司前来投资，其运输条件优良、人才聚集、人文环境独特，也称为"印度的硅谷"。

然而，在 2005 年，这个有着 680 万人口的城市，却只有 2 700 间客房和套间，而同一时期，只有 1 270 万人口的东京就有 87 000 间客房。每年有几百万人来到邦加罗尔，所以短缺的房源使印度的酒店成为"费用最高的酒店"。最高级的里拉皇宫酒店的费用是每间房 390～450 美元一晚，而且 90% 以上的客房和套间一整年都没有空闲。

如果你公司 IBM 的员工也常去印度出差，每年 300 人次，你如何说服对方与你公司签订年度住宿协议？

背景资料来源：《国际商务谈判》，窦然主编，第 100 页。

---

**与酒店经理谈判的开局陈述**

很高兴与 A 先生进行本次对双方公司都很有意义的谈判。本次谈判的目标是与贵公司签订一份 300 人次的年度住宿协议，为 IBM 公司的员工提供差旅方便。

众所周知，邦加罗尔又称"印度的硅谷"，正处于高速发展时期。而 IBM 作为全球最大的信息技术公司之一，在软件和硬件方面都处于业界的领先水平。邦加罗尔的发展与 IBM 的技术和服务是密不可分的。

当然，我们也很清楚，贵公司的房源非常紧张，但办法是想出来的。在合作方式上我们还可以进一步地沟通。我们希望贵公司能从双赢的角度，与我们仔细探讨合作的细节问题，不知意下如何？

---

**情景 3：产品介绍**

注意：在此情景中，全班 6 个小组都以卖方的身份完成产品介绍。

内容：可从产品特点、价格走势、市场优势、未来市场蓝图等内容中选择，也可以根据查找到的资料选择别的思路。

注意事项：

1）产品介绍的内容应经过筛选。第一，对众所周知的产品的一般特性不要使用太大篇幅，毕竟对方对产品也是非常熟悉和专业的。如在铁矿石谈判中，若对铁矿石种类、所含元素、用途等介绍过于详细的话，就会本末倒置。第二，应有重点内容，而不是将所有内容篇幅平均分配。第三，可以结合报价单（有的小组自行准备了一份虚拟的报价单，介绍时更自然）的内容进行介绍。第四，应明确或暗示性地表明与其他供应商的对比优势。第五，应有一些与对方互动的语言，让对方有参与感。

2）在开始介绍时，应将产品的特点分类。如："下面我将从三个方面介绍产品的特点，即……"每种产品都有很多特点，没有分类就很难找到重点。

3）产品介绍应与市场和需求联系起来。产品具有这样或那样的特点固然很好，但不与市场和需求联系起来，对方不一定会放在心上，也不一定能意识到该特点的价值。如："当今社会对于对开门冰箱的需求日增月涨，而我们的冰箱在制造上采用的是外观和质量相结合、销售和服务相统一的原则。我们的设计师通过法国现代建筑的视觉创意启发灵感，在风格上强调实用与百搭的特性，也体现出一种时尚与温暖，这与法国消费者近期的审美观和消费是一致的。当然您也可以多多对比其他供应商的冰箱，就会更明白为什么要选择我们了。"

4）产品介绍时应准确定位己方的身份和工作职责。在国际贸易中，买卖双方的分工和职责与国内贸易差异很大。在模拟环境中，有的同学一下子没有分清楚这样的差异，导致发言中对服务内容等问题分工错位。如有一个冰箱出口商说："在售后服务方面，我公司已实现 2 小时以内保修的上门服务。"这是否能作为出口商的优势呢？在国际贸易中，冰箱在当地市场的售后服务应由买方负责，或者由双方协商负责。卖方可将此售后标准作为一个目标或建议来商讨与对方的合作。

5）对产品的重点优势不要平铺直叙，应有一些相应的语言对此进行强调或美化，以便能引起对方的兴趣，给对方留下深刻的印象。请对比以下两句话。"本款冰箱已通过××认证，全球冰箱唯一的行业认证。""本款冰箱已通过××认证，全球冰箱唯一的行业认证。这是我公司辛勤努力的结果，也是我们的骄傲。"

## 二、磋商阶段

谈判的磋商阶段从报价开始，经历还价、再报价、再还价、僵局等情景。磋商阶段是商务谈判的核心环节，是谈判双方求同存异、协商确定交易条件的过程，也是双方在谈判准备、谈判实力、谈判经验等方面的较量。因此，同学们需要在发言中考虑策略因素。本阶段的实训包括还价和僵局处理两个情景。

**情景 4：还价**

还价又称还盘，是受盘人对发盘内容不完全同意而提出修改或变更的表示，是对发盘条件进行添加、限制或更改的答复。

实训顺序：买方先做，即先由买方根据题目提出折扣要求，再由卖方根据题目要求还价。

参考角度：吹毛求疵、用户需求、负面新闻等，但不限于以上角度。

注意事项：还价时不能直接一句话还个价了事，需要摆出理由，说服对方；发言的形式也不限，可以创造一种自然的气氛，也可以添加高兴、生气、谦卑等情绪。

**买方题目**

背景：卖方报价：离岸价 25 美元/吨，到岸价青岛港 32 美元/吨。

要求：请就该报价提出 8% 的折扣要求。

**卖方题目**

背景：买方就己方报价（离岸价 25 美元/吨，到岸价青岛港 32 美元/吨）提出 8% 折扣要求。

要求：回复最多只能同意 2% 的折扣。

**情景 5：僵局处理**

谈判僵局是指在商务谈判过程中，当双方对所谈问题的利益要求差距较大，各方又都不肯做出让步，导致双方因暂时不可调和的矛盾而形成对峙，进而使谈判呈现一种不进不退的僵持局面。僵局是谈判中常出现的一种局面，而相当比例的僵局是因为沟通不足、偏见、对市场的理解不透彻等主观因素造成的，因此，有效的沟通是解决僵局的重要途径。

注意：说服的途径、形式、内容不限。小组内其他同学可以在表情、动作上营造僵局的气氛。

**买方题目**

背景：你方的底线是 58 美元/吨到岸价，但对方坚持 60 美元/吨到岸价，谈了多个回合，你很气恼。

要求：请使用有效的方式说服对方。

**卖方题目**

背景：你方的底线是 60 美元/吨到岸价，但对方坚持 58 美元/吨到岸价，谈了多个回合，你很气恼。

要求：请使用有效的方式说服对方。

## 三、成交阶段

成交是指谈判各方就所磋商的问题达成共识或意见、观点趋于一致。虽然达成共识，但双方心里仍有可能存在各种疑虑，如该条件是否为自己所能取得的最好成绩、是否还有进一步探索的空间、双方的利益是否均衡、对方的态度是否肯定、是否期望维持良好的合作状态等。因此，还需要做出适宜的表态，以便满足对方的心理需求、建立有利于双方长期合作的基础，使谈判的效益达到最优化。本阶段设置了总结陈词一个情景。

**情景6：总结陈词**

内容：经过半个月的谈判，双方在价格、运输、发货时间、付款方式等方面达成一致。请总结陈词，并表达美好愿望。

注意事项：

1）本情景练习主要是表达对双方努力的肯定和对未来的展望，对价格等交易条件应点到为止，不需要过于具体。

2）成交在望，可考虑增加调节气氛的内容。

# 模拟谈判方案展示（结合 PPT）

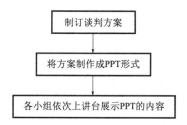

制订谈判方案

将方案制作成PPT形式

各小组依次上讲台展示PPT的内容

## 实训说明

谈判主题：继续使用第一章选择的谈判主题。

在模拟谈判的准备阶段，每个小组应针对本组谈判主题的市场情况以及谈判对手、双方的竞争对手等方面的材料进行搜索，经过归纳总结之后做成一份谈判方案 PPT，并且召开一次内部会议，在小组内部对谈判方案进行展示。

模拟谈判方案展示步骤：每个小组先做好一份谈判方案 PPT，然后派一位同学以主谈人的身份上台，以召开内部会议的形式，将方案介绍给组内其他成员。准备时间为 4 ~ 5 个课时。

---

**提 示**

谈判方案展示的效果由方案内容、PPT 制作效果，及演讲者的表现三部分构成。

其中，方案内容是小组进行进一步谈判准备的基础，对谈判的结果有很大的影响。难度比第二章有所提高。

PPT 制作效果，会直接关系到演讲者的发挥以及听者的理解。

而演讲者的表现，则是整个方案的灵魂。好的演讲者，能把制作一般的内容讲得很精彩。能力一般的演讲者，即使面对制作精良的内容，也很难把其中的内容表达得淋漓尽致。要提高与谈判有关的各方面能力，就要多做相关的练习。

**主要内容和要求** ///

### 一、方案主要内容

模拟谈判方案的主要内容有如下几个方面。

1. 双方的实力对比；

2. 双方优劣势；

3. 产品介绍；

4. 报价（包括贸易术语，结合近期走势）；

5. 市场行情；

6. 人员分工；

7. 风险分析；

8. 谈判策略等。

除了前文提示的问题以外，还应在方案中考虑以下内容。

1. 在利益和关系中，如何侧重？

2. 谈判议题的顺序和时间安排。

3. 人员分工，包括议题分工、策略的运用等。个人风格有异，有能言善辩型的、急切型的、稳重型的、友好型的、严肃型的、善于记录和总结型的、善于发现问题型的、善于打圆场型的、善于建立关系型的等，应针对不同的个性特点做一些不同的安排。

4. 方案中提及的一些内容属于内部机密，但是也要在全班同学面前展示出来。因此像报价这些内容，可以稍作调整。注意：方案中的内容和顺序可自行安排，不应按照本书的提示一个一个地列出来，而应有重点和非重点之分；如有必要，有的内容可以省略。

### 二、篇幅安排

每个组的展示时间为 15 分钟左右。其中应有 1~2 个重点内容，不应将所有内容篇幅平均化。

### 三、提炼内容

方案中的每一个内容都应该是经过提炼的、能与谈判直接联系的，不应该将资料直接切入而不知其说明了什么。请对比这两句话。"B 集团的市场业务范围广，资金雄厚，2015 年公司实现营业收入 1 637.9 亿元。""支付能力是卖方最关注的问题之一。B 集团 2015 年实现营业收入 1 637.9 亿元。"

**发言的形式** ///

作为主谈人，你应以一种了然于胸的姿态进行介绍（脱稿）。发言的形式、风格不限，可自由发挥。可以由一位同学从头到尾自己发言，也可以设置一些互动环节。

发言要求用"讲"的形式，应避免使用"口水话"。如"我们都知道，只有知己知彼方能掌握谈判走向""在了解了我们的优势后，我希望大家不要掉以轻心"等。

发言人在上台前必须预先演练一遍以上。因为作为学生，只有少数人能做到在没有准备的情况下对着PPT的概括性文字就会出口成章、头头是道、自圆其说，大多数同学还是需要经过大量练习的。另外，应演练PPT所有页面的内容，若只演练前面几页PPT，后面发言时效果会有明显差异。

### 语言要求

（一）发言时不可一直自顾自地说，应给听者一些消化的空间

请对比以下两种表达。

1. "以退为进策略。让对方先开口说话，先摸清对方的意图，以退为进，根据对方的条件……" vs "现在说到以退为进策略，这个策略的重点是让对方先开口说话，先摸清对方的意图。注意了，千万不要先说，不要急，有想法也要憋着。明白吗？做得到吗？"

2. "红白脸策略。张小兰负责红脸，李进负责白脸。" vs "现在我们来讨论一下红白脸策略。所谓的红白脸，大家都知道……张小兰，你的笑很甜美，这个任务交给你了。李进平时就挺威严的，你做白脸当仁不让啊……大家都清楚自己的任务了吗？"

（二）应有鼓舞士气的话

如"我想我们每一个人都应该用最热烈的掌声来感谢公司领导们对我们能力的认可。"

### 风险分析

风险是包括多方面的，如贸易术语和支付方式的应用、对方的资信情况、行业风险、宏观环境影响、法律风险、政治风险、文化风险等。

例如，在贸易术语的选择中，需要考虑汇率问题、运输问题、交货地点问题等。如果人民币对美元处于涨势，对于国外这一方来说，在支付国际运费方面并没有太大的影响，但是，对于中方来说，一段时间之后支付国际运费，就可以节省一笔费用。但总的来说，应尽量争取由己方安排运输。对于出口合同而言，如果使用FOB术语结算，进口商和船公司勾结、无单放货的风险很大。另外，外商指定船公司、使用境外货代或无船承运人的情况与日俱增。但很多出口商不了解中国的相关规定，也没有认真对境外货代或无船承运人的资格进行审查。事实上，中华人民共和国交通运输部规定，境外货代提单必须委托经我国有关部门批准的货代企业签发，货主可要求代理签发提单的货代企业出具在目的港凭正本提单放货的保函。同时也规定，经营无船承运人业务，应当向国务院交通运输主管部门办理提单登记，并缴纳保证金。不符合以上条件的境外货代和无船承运人的任何操作，均属违法经营。对于进口商而言，使用C组贸易术语，如果是因为船公司的责任耽误了交货，进口商则没有资格向船公司主张权利。

再举个定牌生产的例子。出口商比较草率，得到知名品牌的委托，就只签订简单的合同。事实上这样做的风险很大。虽然知名品牌一般不会违约，但也会出现经营困难甚至财务危机导致无法支付的情况。这样的例子是真实发生过的。在这种情况下，出口商将面临什么局面呢？如果向法院提出诉讼，即使胜诉，对方公司也没有能力继续履行合同或承担赔偿责任；如果在对方公司未注册商标的国家或地区处理货物，第一，会发生很大的折价损失；第二，中国业已加入《关于商标保护的马德里公约》，出口商将面临侵权责任。那么，风险应该如何避免？最好的方法是签订详尽的合同。国外知名产品的定牌销售协议往往有厚厚的几十页，从设备到工艺，从标准到质量，从配件到整机，从价格到市场，从服务到维修，从交

货周期到配件供应年限，从付款到交货，从检测到验收，以及商标权的问题，均应提前进行详尽、明确的约定。

注意：在本环节中，每个方案中仅需要做一个风险分析即可。

## PPT 页面要求

### 一、分析 PPT 在内容和画面方面存在的问题

图 4-1 所示为 B 集团优势 PPT，其存在问题分析如下。

**B集团优势**

- 规模优势：B集团是国内最大、世界前三的钢铁企业；业务规模涵盖钢铁冶炼、加工、电力、煤炭、工业气体生产、码头、仓储、运输等与钢铁相关行业。中国重要的钢铁企业，效益好，资金实力雄厚
- 所在国是全球最大的铁矿石需求国，对铁矿石定价有较大的影响。
- 科技水平发达，对矿石提纯度提高具有促进作用。
- 其他买家的供给。和国内、国外多家铁矿石企业有合作协议，且多家是必拓的竞争对手，谈判余地大。

**图 4-1　B 集团优势 PPT**

1. 文字太多，使听者找不到重点；发言者容易进入"念"的状态。
2. 内容与谈判的联系不够紧密，如业务规模、科技水平与铁矿石关系不大。
3. 内容深度不够，如提到铁矿石需求，应有数据作为支撑；如提到其他买家，应有名称和近几年的交易量。

### 二、好的 PPT 画面示范

好的 PPT 画面应做到简洁、概括性强、重点突出，有利于发言者讲解，图 4-2 所示的支付能力 PPT 可作参考。

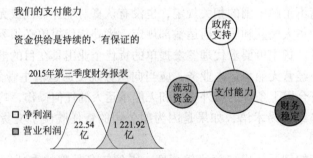

**图 4-2　支付能力 PPT**

## 谈判方案展示范例

图 4-3 所示为我方利益及实力 PPT。

就实力来说，我方算是综合实力比较强的。作为法国电器行业的优质领导者之一，我方资金比较雄厚，在多年的合作里，建立了比较牢靠的市场基础。在了解我方的实力之后，希

**图 4-3　我方利益及实力 PPT**

望大家能在谈判的过程中，有一个比较自信的依靠感。在明确了我方的实力和利益之后，我们再来看看谈判对象，即中国 H 集团，它的利益及实力。

图 4-4 所示为对方利益及实力 PPT。

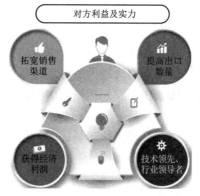

**图 4-4　对方利益及实力 PPT**

中国 H 集团如果能成功与我方进行这一次的合作，也能拓宽它的销售渠道，提高它本身的出口数量，获取经济利润。另外，中国 H 集团的实力也较强，因为它有一个优质的、一体化的营销渠道，技术也非常好，也算是一个冰箱品牌市场的行业领导者。

图 4-5 所示为双方优劣势分析（我方优势）PPT。

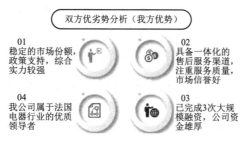

**图 4-5　我方优势分析 PPT**

接下来是关于双方的优劣势分析。

在了解我方综合利益和实力的基础上，再明确我方在本次谈判中所处的地位以及我方的优势。第一，我公司有一个稳定的市场份额，基于本土企业的快速发展，也得到政府政策支持的这样一个依靠，所以可以说，综合实力也是比较强的。第二，我公司具备一体化的售后服务渠道。我公司已上市，注重服务质量，市场信誉还是比较好的。第三，在过去的几年中，我公司完成了 3 次大规模的融资，在资金这一块，还是比较充裕的。第四，我公司是法国电器行业的优质领导者，这是毋庸置疑的。

基于以上几点，我相信，我公司的优势已经非常明了了。

当然，我们说，有优点同时也有不足，人非圣贤，谁也做不到十全十美。认识到优势之后，也要意识到我们的劣势和不足。图4-6所示为双方优劣势分析（我方劣势）PPT。

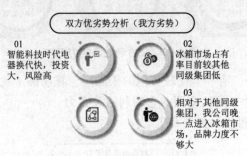

**图4-6 我方劣势分析PPT**

首先，在智能科技时代技术快速更新的背景下，电器的更新换代非常快。对于冰箱出口的新产品来说，我们做进口销售，投资是比较大的，风险也比较高。

另外，目前相对其他同级集团来说，我们冰箱市场的占有率不高，还没有处于主导的地位，可以说，尚处于中下水平。

最后，相对其他的同级集团，我公司进入市场的时间稍晚，品牌的力度还不够强大。

认识到我方的劣势之后，我们来了解本次谈判的对方，中国H集团。图4-7所示为双方优势分析（对方优势分析）PPT。

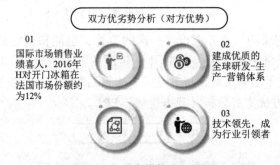

**图4-7 对方优势分析PPT**

第一，我们获取的资料显示，对方在国际市场的销售业绩非常喜人。2016年，对方的H对开门冰箱在法国的市场份额约为12%。在其他竞争品牌共同来开拓法国市场的前提下，这个数据还是比较有竞争力的。

第二，中国H集团建成了一个优质的全球研发—生产—营销体系，可以说一体化性质还是非常强的。

第三，中国H集团的技术比较领先，算是行业的领导者。当然，根据收到的资料，我们知道，对方也是有劣势的。

我们可以了解对方的劣势，以此作为谈判的突破点之一。图4-8所示为双方劣势分析（对方劣势分析）PPT。

第一，他们对法国市场的了解还不够完善，可以说消费者对H对开门冰箱的这两款新产品还是完全不熟悉的。我们的投入对他们来说也是拓宽渠道的一个非常迫切的需求。

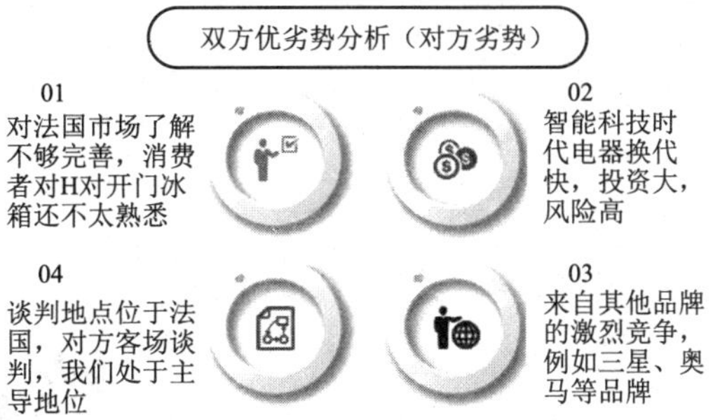

**图 4 - 8　对方劣势分析 PPT**

第二，和我们的劣势一样，在智能科技时代，电器的更新换代非常快，投资大，风险也非常高。

第三，相对于其他品牌来说，他们的市场份额虽然有 12%，但是也不缺少其他品牌的激烈竞争，如韩国的三星和中国的奥马等。

第四，他们劣势的主要原因是，谈判的地点在法国。对我方来说这是本土谈判，我方能处于一个主导的地位。

资料来源：广西财经学院国际商务 1621 班，演讲者：黄青妙。

# 综合模拟谈判

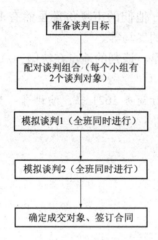

准备谈判目标

↓

配对谈判组合（每个小组有
2个谈判对象）

↓

模拟谈判1（全班同时进行）

↓

模拟谈判2（全班同时进行）

↓

确定成交对象、签订合同

## 准备谈判目标

### 一、时间安排

约 3 个课时用来准备谈判目标的内容。另外，小组内部用 1 个课时对谈判目标的内容进行统一、熟悉和掌握，要争取做到所有成员对每一个细节都能达到统一，避免出现谈判时成员之间立场不同甚至相互矛盾的情况。

### 二、内容

谈判内容有成交价格、贸易术语、成交数量、支付方式、交货日期、运输方式、售后服务等。

根据谈判方案、书后附录的国际贸易合同、市场调查和技术优势等内容，讨论和制订模拟谈判的具体目标。制订谈判目标应注意，成交价格与贸易术语、成交数量、支付方式、售后服务等方面因素是挂钩的，若这些因素有所改变或者存在调整的空间，那么应列出在不同情况下的成交价格。

### 三、调整项目

谈判目标应包括可让步的项目和不可让步的项目，如贸易术语、包装、发货时间、付款方式、成交数量、质量标准、检验方式、售后服务等，以便谈判时谈判人员能够明确在哪些方面应该坚定，哪些方面可以再讨论和协调。

### 四、选择使用特殊条款

1. 交货期确认，即规定交货期，若产生延期，卖方必须赔偿一定金额。

2. 质量条款确认，即要求质量是某个参数，如果低于这个参数，买方可以退货或要求赔偿。如在铁矿石销售合同中有"如果铁含量每超过基本铁含量（62.5%）1%，则每吨价格应上调＿＿＿＿＿美元。如果铁含量每低于基本铁含量（62.5%）1%，则每吨价格应下调＿＿＿＿＿美元"这类的条款，那么在谈判目标中就应该包含这一类的内容。

3. 交货标准确认，是按照质检单的标准交货还是按照样品的标准交货，确认后，若交货不满足要求，买方可以要求退换货。

4. 单据制作确认，即规定如果因为卖方制作单据出现错误影响买方提货，卖方应承担一切责任。

5. 包装条款确认，即包装使用标准、包装贴物等问题，如果出现不一致，买方可索赔一定数量的金额。

6. 在售后服务方面，应考虑对方可能提出的不同要求。如买方要求卖方在买方国家长期派驻多名技术人员，那么技术人员的工资和各项费用应如何解决？双方各承担一部分还是某一方单独承担？若双方各承担一部分，那么应该分别承担多少？……

以上这些特殊条款，若能同意写进合同之中，则可能成为拿单的利器。但是，也需要加上一些限定条件，最大限度保障我们的利益。例如，在交货期方面，可以加上"若因政治、极端天气等不可抗力的因素引起延误，双方本着谅解的态度商定新的交货期。"在质量条款方面，可以加上"检验方法为……，不接受其他的检验方法得到的结果。"在单据制作方面，可以加上"如果因为买方未及时提供信息，或者提供了错误信息，或者完全按照买方意思制作的单据出问题，我方概不负责。"在包装条款方面，可以加上"我方会在交货时拍照，如果交付时完好，符合要求，而在运输途中出现的问题，我方概不负责。"

*资料来源：《JAC 外贸工具书——JAC 和他的外贸故事》，经编者整理改编。*

**配对谈判组合** ▶▶▶

首先，一个买方小组和一个卖方小组的组合，即一个谈判组合。

其次，每个小组分别和两个谈判对象完成谈判之后，再确定和哪一方签订合同。买方和卖方都处于双向选择的位置。最终的结果是，可能有的小组能签下两单，而有的小组一单也拿不到。这样的结果，考验的不仅仅是实力，还有准备程度、双方的契合程度、实际的发挥等因素。这与实际的国际商务谈判是很相似的。因此，还需要同学们做好充足的准备，争取至少能拿到一个订单。

## 综合模拟谈判

### 一、时间安排

其中，每个小组将先后与两个谈判对象完成谈判，与每个谈判对象的谈判时间为4个课时。全班同时进行谈判时，4个课时之内完成第一次谈判。之后交换谈判对象，再进行4个课时的谈判。

### 二、模拟谈判情景设置

谈判的地点设置为国外买方/卖方的地点，即中方为客场谈判。模拟谈判的情景设置如下。

1. 谈判开始情景：模拟谈判从双方都已进入会议室，在谈判桌旁握手开始。

2. 谈判流程如下。

1）主方主持：主方应有一个主持人，欢迎来宾、介绍公司成员、简单阐述谈判目标和表达美好愿望。注意致欢迎辞之前先自我介绍（职位、名字）。

2）客方发言：客方有一个代表发言，答谢、介绍公司成员和阐述谈判目标、表达美好愿望。注意答谢之前先进行自我介绍（职位、名字）。

3）正式谈判：以上环节完成之后进入开局陈述、报价、还价阶段，产品介绍可以自由选择在某个环节进行。不规定哪一方先报价，由各个小组根据自己的策略决定先报价与否。

4）记录：在谈判的过程中，每个小组应有一位同学专门记录谈判内容。记录的重点不是每个人说了什么，而是谈判的议题和线索，即第一个议题是什么，是哪一方提出的，另一方应如何回应，哪一方胜出，己方是否还有研究的空间，是否可以在下一轮谈判中作为筹码使用，等等。记录很重要，这是谈判团队在激烈的辩论之后理清思路，制订下一轮谈判计划的关键。因此，记录员需要有敏锐的洞察力、严谨的判断力和敏捷的记录能力。

5）休会：谈判有阶段性进展或陷入僵局之后应休会，重新调整思路、查找资料、总结并制订下一步计划。谈判过程中不需要根据上课时间停止或继续谈判，谈判的休会、重启等环节应根据谈判的内容和进程而定。休会期间可以去旁听其他组的谈判。

6）其他：在模拟谈判的过程中需要使用大量的辅助材料，并需要做记录，因此模拟谈判过程中不需要脱稿。

### 三、确定成交对象、签订合同

时间：2个课时。

两次谈判完成之后，每个小组从两个谈判对象中选择成交的对象，然后双方签订一份国际贸易销售合同。合同签订好之后交给老师。

没有拿到订单的小组，需要写一份报告，说明意向与之签订合同的小组，并将双方谈判的结果参照合同的形式一一列出（包括达成一致的内容和未达成一致的内容）。

## 谈判技巧

### 一、回答问题的步骤

回答问题的步骤为"认同—但是—表示"。

请参考第二章"个人快速反应练习"的回答步骤及范例。

## 二、开局气氛的营造

在理论课中已经学习过,开局包括自然和友好的开局、冷漠和压抑的开局、高调开局、拖延开局、敷衍开局等。在大多数情况下,会采用自然和友好的开局,只有在特殊情况下,如对方给的压力太大,或者我方需要展示特殊立场时,才会使用其他种类的开局。

开局气氛的营造,往往就在于正式谈判前的五分钟"闲聊"时间,即使采用自然和友好的开局,话题的选择也会影响营造的气氛。在一次课堂练习中,老师让同学们分组思考:当你的客户分别来自不同的国家,这五分钟你将和对方聊什么,怎么聊呢?同学们的回答中,在内容上有关于天气、体育和生活的,也有关于经济(从经济学的角度)、政治和宗教的;在表达方式上,有赞赏对方的,也有和中国做比较的。那么,在这里,要特别说明以下两点。

在内容上,聊天气、体育和运动应是很好的选择,因为内容比较简单,容易引起共鸣,也能聊得下去;但若从经济学的角度来聊两国的发展,就过难了。毕竟大家都是商人,不是经济学家,如此高深的话题,大概率的结果是冷场。而政治和宗教,都是应该避免的话题,有很多禁忌之处,也容易产生分歧。

在表达方式上,以赞赏的方式提到对方的优势是非常好的,这样大家都很高兴,也容易形成共鸣、建立信任或者相互欣赏。例如,有的同学跟德国客商聊德国的建筑风格,跟俄罗斯客商聊《钢铁是怎样炼成的》,跟法国客商聊红酒,这样客商的兴致会很高。但是,有一个小组和英国的客商说:"这几天天气很好,阳光明媚。听说伦敦是雾都,您觉得哪里的天气更好呢?"这个说法有什么问题呢?首先,我们聊天的目的,是建立共鸣,而比较是很容易导致分歧。其次,这次比较,是把对方比下去了,人家会高兴吗?谈判还未开局对方心里就不好受了,对后续谈判尤为不利。

---

**以下说法有什么不合适的地方?**

有一个小组问沙特阿拉伯的客户:"您觉得'一带一路'给你们公司带来了什么好处?"请问这个说法有什么不合适的地方?

第一,在内容上,过于宏观和抽象,对方需要思考和总结,冷场出现的可能性很高,话题难以轻松继续。

第二,这个说法是带有优越感的,容易给对方带来不快。中国正在快速崛起,我们在和外国客户交流时,很容易不经意展现出优越感。如果不是为了给对方压力而特意营造如此气氛,那么有关优越感的内容应该尽量避免,使谈判和合作在平等的氛围下进行。

---

## 三、报价策略

### (一)报价时要考虑是否要先报价

虽然一般是卖方先行报价,但这并不是一个规则。先报价和后报价各有优劣。先报价的一方,可以为谈判设置一个最高价或一个最低价,也能表现出己方的风格和态度;但是,如果先报价的一方是新手,对市场的行情不是很了解,也有可能报出一个过低或过高的价格,要么暴露出自己的无知,要么让对方觉得你毫无诚意。后报价的一方,可以揣摩对方的立场和态度，后发制人，但也失去了掌控局面的时机。

### （二）报价时需要考虑心理因素

1. 除法报价法。若总价的数额比较大，可以考虑将总价以某个单位进行拆分，较小的单价会比较让人接受。例如，一台高端冷暖空调的报价为899美元，如果我们说明，空调的使用寿命是12年，即4 380天，那么每天的花费仅有0.2美元，也就是说，仅需十分之一包薯条的价格，就可以享受夏日的清凉舒爽和冬日的温暖舒适。如果对方说，你们的空调虽好，但比别人的贵了80美元一台，那么我们就可以说，每天仅多了0.01美元，不到一根薯条的钱，却可以享受更好的质量和更多的功能，这是完全值得的。

2. 尾数报价法。这是生活中常见的一种报价法，即通过尾数的处理，使价格呈现出一种更便宜或更昂贵的感觉。例如，对于普通的物品，如果价格是1.99元、9.9元、99元，会让人感觉比2元、10元、100元便宜了一个档次；对于奢侈品而言，20 000元与19 999元之间的心理差异，也是远远大于1元的，能让人感觉很划算，因为档次高了一级，价格却没有贵很多。

3. 加法报价法。有的商品可以成套出售，也可以单件出售。成套物品的价格较高，如果直接推荐，可能会超出对方的预期，成功率就不高。但是，如果在对方可能愿意接受的范围内，先将其中的一两件推荐出去，再慢慢地增加其他的物件。当愿意购买的物件接近整套时，对方很有可能咬咬牙就一起买了下来。

4. 积极价格和消极价格。每种商品或服务都会涉及众多的特点和参数，那么，我们在介绍时选择哪个点来介绍，才能引起对方的共鸣或兴趣，这是需要准备和研究的，也可能需要一些运气。例如，化工品有纯度、水分含量、杂质含量、重金属含量、游离甲醛含量等指标，不同的客户关注的指标不同，这与他们如何使用产品有很大的关系。能将对方在意的特点呈现出来，让对方感兴趣，这就是积极价格。相反，相同的商品，相同的价格，如果我们介绍的特点不是对方感兴趣的，对方就会觉得这个商品或服务是不值得购买的，这就是消极价格。举个化工产品生产企业与外商谈判的例子。这个企业的产品在水分、灰度方面与同行基本一致，但颗粒稍微粗一些（溶解度更差），价格也没有优势。但是企业的谈判人员很注重员工的使用体验（同行很少关注），即员工在使用过程中危险发生的概率，刚好这一点也是外商目前所关注的，双方产生了很大的共鸣，订单就顺利拿到了。

5. 整数策略。谈判进入尾声，成交在即，可是对方还有些犹豫。如果你是卖方，产品单价是269美元/箱，对方的订购数量是60箱，即总价是16 410美元。如果你方主动提出取整数，即成交金额为16 000美元，是否可以加速成交呢？成功的可能性还是很高的。

## 四、具备一定的专业素质和能力

在选择合作伙伴的时候，会考虑很多因素，包括对方和对方公司是否可靠、对方的实力和专业性、对方的合作态度、对方对未来发展的关注点等。

那么，专业能力从哪些方面体现？这就需要深入地了解产品的产业链上中下游结构，包括原材料、制作工艺、管理和贸易，以及技术指标、性能、外观、优势、应用、认证、创新、进出口特点和案例、国际市场动态等细节。以纺织服装为例，专业的人员应了解以下几个方面的知识。

第一，纺织服装行业的产业链结构和内容，如表5-1所示。

表 5-1　纺织服装行业产业链

| 环节 | 大类 | 细节 | 影响因素 |
|---|---|---|---|
| 上游 | 天然纤维 | 棉 | 企业设备的自动化能力 技术研发能力 成本和规模 订单稳定性 |
| | | 麻 | |
| | | 丝 | |
| | | 毛 | |
| | 化学纤维 | 再生纤维 | |
| | | 合成纤维 | |
| 中游 | 纺织过程 | 纱线加工成坯布 | |
| | | 坯布印染或印花成面料 | |
| | 服装成品生产 | — | |
| 下游 | 服装管理及贸易环节 | — | 品牌 产品设计 渠道 供应链 |

第二，面料、服装的主要技术指标。纺织品面料的主要技术指标包括长度指标、经纬密度、幅宽、克重；服装的主要技术指标包括织物缩水率、织物缝纫强力、织物断裂强力、织物撕破强力、织物色差等。

第三，服装的分类和等级。根据 GB 18401-2010《国家纺织产品基本安全技术规范》的规定，服装有 A 类、B 类和 C 类。A 类是 36 个月以下的婴幼儿用品；B 类是直接接触皮肤的产品；C 类是非直接接触皮肤的产品。另外，服装以件为单位，根据外观、规格、色差、疵点、缝制等指标，分为一等、二等和三等，便于双方贸易签约时使用。美国《联邦条例法典》对所有纺织物根据燃烧时间的长短，分为 1 级、2 级和 3 级，即正常燃烧性、中等燃烧性和快速燃烧性。

选择服装主题的同学，需要根据以上基本内容，对所涉及的产品进行深入的了解，才可以具备一定的专业素质和能力。选择其他主题的同学，也可以借鉴以上的框架和思路。

---

**英语不好，但业绩第一**

我们公司有一个小姑娘，英语不好，连基本的沟通都很勉强。但是，她的业绩却是我们公司的第一名。为什么？

素材！

她有接近 20GB 的产品图片和视频资料，而其他人不到 3GB，这些还是他们入职时公司给的。

小姑娘用入职后第一个月的工资买了一台数码相机，每每发货，必须去拍。她问工人这个设备哪里做得最成功，哪里返工了，为什么返工，哪里独特，然后一一拍照，再去分类整理、记录。

跟客户聊天，她动不动就是图片、视频，对产品的各种特点、使用情况都了如指掌，你有吗？

资料来源：《JAC 外贸工具书——JAC 和他的外贸故事》，经编者整理改编。

### 五、横向谈判与纵向谈判

横向谈判是指在确定谈判所涉及的主要问题后，开始逐个讨论预先确定的问题，在某一问题上出现矛盾或分歧时，就把这一问题放在后面，讨论其他问题。如此周而复始地讨论下去，直到所有内容都谈妥为止。例如，在资金借贷谈判中，谈判内容要涉及货币、金额、利息率、贷款期限、担保、还款以及宽限期等问题，如果双方在贷款期限上不能达成一致意见，就可以把这一问题放在后面，继续讨论担保、还款等问题。当其他问题解决之后，再回过头来讨论这个问题。这种谈判方式的核心是灵活、变通，只要有利于解决问题，经过双方协商同意，讨论的条款就可以随时调整。也可以采用这种方法，把与此有关的要点一起提出来，一起讨论研究，使你所谈的问题相互之间有一个协商、让步的余地，这非常有利于问题的解决。

例如，当对方提出："我们可以和你谈谈，可问题是，我们要在新奥尔良举行年度销售会议，如果希望成为我们的供应商，你们就必须在举行销售会议那个月的1号之前交来样本，否则，我们也就没有必要浪费时间了。"这个要求你们是做不到的，但这未必是对方真正的底线，谈判是否还可以进行下去呢？可以这样说："我们知道这对你们很重要，但我们不妨先把这个问题放一放，讨论一下这项工作的细节问题。比如说，你们希望我们使用工会员工吗？关于付款，你们有什么建议？"

纵向谈判是指在确定谈判的主要问题后，逐个讨论每一问题和条款，讨论一个问题，解决一个问题，一直到谈判结束。例如，在一项产品谈判时，双方确定了价格、质量、运输、保险、索赔等项主要内容后，开始就价格进行磋商。如果价格确定不下来，就不谈其他条款。只有价格谈妥之后，才依次讨论其他问题。

横向谈判和纵向谈判各有优缺点。例如，横向谈判是一种比较灵活的方式，有利于双方对各种问题的解决，但是也有可能增加双方讨价还价的余地，甚至可能会使议题逐渐偏离主线，过多关注一些非关键的问题。纵向谈判可以把复杂的问题简单化，避免多头牵制、议而不决的弊病，但也会导致议程确定过于死板，讨论问题时不能相互通融，不能充分发挥谈判人员的想象力和创造力，也可能会因为某一个陷入僵局的问题而严重影响谈判的进程。

### 六、事实胜于雄辩

我们在与对方讨价还价时，必须以事实为依据。例如，在技术引进的谈判中，我们要求对方降价，应列出降价的理由。例如，这几项技术，有几项是新研制的，几项是即将过期的？对方公司每年在研发方面的投入大概是多少，成果有多少项，平均到每项技术大概是多少？在同行中的参考价格是多少？……

### 七、通过微表情和身体语言等信息判断对方的关注点和决定

在沟通的过程中，尤其是在发现对方一直没有进入状态的时候，寻找切入点非常重要。要想合作，必须先有共鸣。那么，如何才能找到对方关注的话题呢？对方表情或动作的明显变化，则是一个重要的信号。

## 八、应避免的话题或说法

### （一）直接表明这是公司规定、行业规则

例如，在付款方式上，如果直接告诉客户："我们公司一向只接受不可撤销即期 L/C 付款方式"。对方会怎么想？这个说法本身并没有错，但如果你是对方，你是否觉得还有继续谈判的必要？双方做生意，是为了达成合作、满足各自所需，而不是简单的谁去迁就谁，除非对方已经没有了选择。事实上，当客户对付款方式提出不同的要求时，我们可以考虑对方的如下需求。

1）如果客户更倾向于节省成本，那么 T/T 是更好的方式。

2）如果客户更在意资金的流动性，那么他们就更希望能采用信用证的付款方式。

3）如果客户需要更快的发货时间，那么信用证无疑就不是最好的选择。

当然，我们也要弄清楚一些特别的国家政策和遵守某些行业已形成的固定付款方式。例如：孟加拉国要求所有进口必须是 L/C + CNF，大部分产品都要做第三方检验，SGS 检验或者 BV 检测；在美国，记名提单复印件可以提货，如果你和美国客户做 "T/T 30% 预付款，剩余 70% 见副本付款" 就会有很大的风险；土耳其、印度等国家规定，转运或退运必须有原购买者的书面证明，滞留港口的货物超过一定日期，就要被拍卖，原购买者有优先拍得权等。在行业规则方面，以机械类和农药类为例，机械类产品的付款方式一般是 30% 预付，70% 到厂验货付清，然后出厂；某些农药，90% 是放账，几十天的信用证甚至是承兑交单或付款交单。

### （二）在国外客户面前用自己本土的语言谈笑风生

使用本土语言这个做法本身并没有错，但是对方会觉得自己不受尊重，也很有可能因为这一时的不高兴而否决我们前面所有的努力。要明白，外贸谈判的选择余地是很大的，每个谈判人员都在不停地选择和被选择。在真正签约前，我们的客户都在对比我们和竞争对手的各个方面，如产品、服务、实力、合作前景甚至接待得好坏等。

### （三）贪口头上的便宜，和客户争一时长短

当客户提出一个根本不可能实现的要求时，你是笑笑，以平和的心态明白，对方只是说说而已，还是纠缠于这个细节，告诉对方这是不可能的，他到哪都拿不到这个条件，等等。说完，你自己心里舒坦了，可客户却不舒坦了。

### （四）言多必失

该说的说，不该说的也说，你把对方本来没有想到的问题也都提了出来。当然，这个做法具有双面性，客户也有可能认为我们比较坦诚，是可以信赖的合作伙伴。但是，如果谈判基本完成、签约意向基本确定时无意中把产品或服务的缺点（尤其是可以忽略的缺点）暴露出来，客户很可能会想，"哎呀，这个问题之前没有想到，让我再回去想想"。然后重新对比一番，这等于重新洗牌，结果功亏一篑。

## 九、谈判座位的安排

长桌的座位安排，如图 5-1 所示。

首先，在主人和客人的座位安排上，应注意让客人坐在面对门口的那一侧，以便给客人

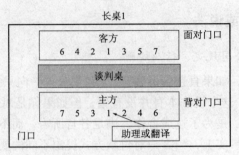

**图5-1　长桌1的座位安排**

带来安全的感觉。这是主方对客方尊重和诚意的表示。

其次，在小组内部的座位安排上，主谈人应坐在中间，第二重要的人员坐在主谈人的右边，第三重要的人坐在主谈人的左边，以此类推。如果主谈有助理或翻译，可以安排在主谈的右后方。

图5-2所示为长桌2的座位安排。

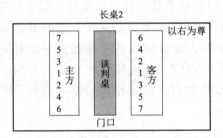

**图5-2　长桌2的座位安排**

在这种类型的谈判桌上，应遵循"以右为尊"的原则，让客人坐在谈判桌的右侧，以表示对客人的尊重。小组内部的座位安排同长桌1。

圆桌的座位安排如图5-3所示。

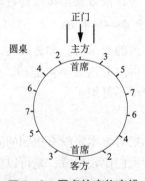

**图5-3　圆桌的座位安排**

圆桌谈判的座位安排与长桌1的安排相同，应注意让客方坐在面对门口的那一侧。

小组内部的座位安排同长桌1。

穿插的座位安排：双方各坐一方，便于小组内部的交流和团结，但同时也会产生一种对抗的氛围。而穿插的座位，可以极大程度地缓解这种对抗。

---

**坐错位置也行？**

美国著名的谈判权威尼尔伦伯格有一次被邀请去参加工会与管理人之间的谈判。他作为管理人一方的谈判代表。在被介绍之后，工会代表请尼尔伦伯格坐在他们的对面。但是，尼尔伦伯格却与工会代表同坐在一起。工会代表们都用奇怪的眼光看着尼尔伦伯格，示意他坐错了位置，可尼尔伦伯格却不予理会。

谈判开始不久，工会代表几乎忘记了尼尔伦伯格是代表管理者的一方，他们仔细倾听了他的分析、意见和建议，就像是在听自己一方的意见和建议一样。对抗的气氛锐减，而融洽的气氛陡增。尼尔伦伯格对谈判座位的选择，可以说为这次谈判的圆满成功助了一臂之力。

---

## 谈判中可能会遇到的困境及解决方法

### 一、对方采用影子策略

对方说有一个竞争者（B 公司）提供了更优惠的条件，其中会存在三种情况。

1. 知道 B 公司，而且 B 公司比我们公司实力更强、信誉度更好。

2. 知道 B 公司，但 B 公司的实力和我们公司差不多，或者更差一些。

3. 没听说过 B 公司。

那么，遇上这些情况，我们应该怎么办呢？

首先，在陈述时，可使用"认同—但是—表示"这三个步骤，让对方先从心理上感到认同，然后才会听得进我们的陈述。其次，表达我们以下的看法。

第一种情况，首先要祝贺对方得到大公司的橄榄枝。然后表示，大公司有大公司的特点和规则，可以向对方询问 B 公司的具体条件，再列举反对的理由，如行事僵化、起订量高、付款方式不灵活、发货时间和地点不够合适。甚至可以说，大公司明明可以提出更高的要求，为什么会自降身价，是不是最近有什么丑闻，等等。最后再说明我方的优势，"你看，我们在经营上、战略上有这么多的共鸣，我们的服务至上，我们的原材料和技术工艺和 B 公司是一样的水准"，等等。要注意的是，对方提出 B 公司，也有可能是使用影子策略借机压价，不一定是真的，不要一听到有大公司来竞争就乱了阵脚。

第二种情况，B 公司和我们实力相当，或者更差一些。我们的态度应是，不紧张，但也要重视，而且要坚持立场。首先，认同 B 公司的条件确实不错。但是，也会有缺点的。B 公司和我们实力差不多，或者更差，那么对方看重的是什么？有没有和他们具体谈过？有没有实地去了解过？他们的经营理念是什么？重点市场在哪里？贸易习惯是什么？……让对方对该合作产生犹豫，最后再抛出我们的橄榄枝，列出我们的优点。

第三种情况，我们没有听说过 B 公司。那就请问对方，B 公司的具体情况，是哪个国家的公司？在这个行业经营多少年了？主营业务是什么？经营特色是什么？有没有固定的合作伙伴？……相比之下，我们的合作更有保障，更值得信赖。当然，也不可轻易做出让步。

### 二、对方提出竞争产品功能多

对方提出其他公司的产品，比我们的多了一个功能，当然价格也稍贵。

举个例子，产品为破壁料理机。对方问："你们的产品没有'点动清洗'功能，我们不太满意。其他公司这一款虽然都稍微贵一些，但是有这个功能。"

遇到这种情况怎么办？首先要明确自身的立场。思考两个问题：一是是否便宜没好货，好货不便宜？我们的价格稍低一些，是否质量就差一些？二是我们还有拿单的可能吗？客户是否真的被那个产品吸引住了？

第一个问题：一分钱一分货。我们的产品稍微便宜一些，设计和质量上也会稍微差一些。第二个问题：别人的产品更好，并不表示客户就一定要签别人的单，要不然怎么会来到这里和我们谈判？也许客户在拿两方做比较，也许只是为了压价。在成熟、多样化的市场里，产品有不同的价格和质量是正常的，它们也对应着不同的消费人群。更高端的产品销量不一定更大，价格适宜的产品市场潜力也不一定更小。所以，成交的可能性是有的。

那么，接下来应该怎么做才能促成交易呢？

首先，要恢复对我方产品的信心，包括己方的信心和对方的信心。可以问对方："您认为这类产品，最核心的价值是什么呢？"毕竟两个产品价格相差不大，所以多出来的那个功能，肯定不会是核心功能。在破壁料理机的贸易谈判中，产品的核心是电机、刀头、不锈钢和玻璃等部件，"点动清洗"并非核心功能。

其次，分析那个非现有的功能是否可以通过其他现有的功能替代？这个功能是否非常必要？消费者是否非常在意这个功能？等等。例如，破壁料理机的"点动清洗"功能可以通过"果蔬汁"功能实现，也并非消费者非常在意的功能。

最后，再次确认我方的产品是物有所值的。这就需要对产品和市场进行分析，如性价比、其他特点、认证证书、原材料优势、目标市场、以往销量等。总之，对比了之后，再把对方的注意力转到我们的产品即可。

被对比的情况在谈判中是很常见的，而且对方举出来的例子很可能比我们的更好。这个时候，最重要的，是对自己、对谈判有信心。"不被迷惑—分析产品—分析市场—重拾信心—促成交易"是非常有效的步骤。

### 三、谈判基本完成，但是对方有顾虑

如果是第一次合作，对方担心我们是否能信守承诺、正常履约；我们所报条件合情合理，已经符合成交的要求，可对方总是犹豫不决，不置可否。

对于顾虑，我们要明白，有顾虑是正常的，毕竟国际商务风险很大。但我们是否要说，顾虑很难消除，合作就很难推进了呢？答案是否定的。在顾虑中合作，是国际商务的常态。不论跟谁合作，都是有顾虑的。我们与竞争对手较量的，不仅是公司资产和名誉、产品质量、服务、合作条件等硬实力，还有接待、沟通、合作蓝图等软实力。而顾虑的消除，是沟通中一个基本的内容。信任感是合作的基础。消除顾虑、建立信任，这也是谈判的一个基本任务。

那么，要如何消除顾虑呢？首先，要明确对方的顾虑是什么？其次，再做出合理的解释。推荐思路如下。

1. 我理解你的顾虑。如果是我，我也会有这些顾虑。

2. 我们在意的是……失言对我也没有好处。

3. 合作对我们来说是双赢的。

4. 列举事实和数据（如果有的话）。

## 四、对方准备更充分

对方准备比我们充分，或者气场更足。总之，整个局势基本被对方控制。在实际的谈判中，谈判失去主动性的一方是经常吃亏的，原因就是没有及时醒悟，糊里糊涂地按照对方的条件签了约。

但是，这种情况真的很常见，因为势均力敌的情况不是经常出现的。在这种情况下应该怎么办呢？首先，绝不可以让对方牵着鼻子走。要及时清醒过来，借助休会的时机，重新调整思路。其次，要借助反向思维的方式，跳出对方的圈子。例如，我们要思考，他们说的听起来很有道理。是真的有道理吗？他们说这样不行。是真的不行吗？他们说别人已经占了先机，我们就没有机会了吗？他们说到竞争对手的优势，真有这么厉害。我们是不是就没有一点回旋的余地了呢？……提出问题，而且要有了答案之后，才能再返回谈判桌。

---

**沟通发言技巧**

当你（供应商）问对方："你之前是在哪里采购的？"对方回答："我第一次来中国采购，以前在××国（名字）采购。"你接下来说什么？

第一，接过对方的话茬，首先表示很高兴对方能来中国采购，然后介绍中国在制造业和产业链方面的优势，肯定对方能在中国找到满意的合作伙伴。

第二，咨询对方之前采购的一些细节，可以试探性地询问对方对以前的合作有哪些方面不满意，或者根据对方的描述来做一些对比。

注意：在谈判中，双方沟通应达到三个目的：获取信息、表达信息、建立信任。因而，在沟通时，除了要表达自己的立场和展示准备的材料之外，还要善于询问，以便获得更多的信息，提高沟通的效率。

---

**模拟谈判中常见的问题** ////

### 一、回答前没有进行思考和讨论

回答时不假思索。没有思考、事实和数据的查询和团队的默契，以至于很多有意义的议题只辩论一个回合就没有了下文，也没有进一步的研究和利用，实在可惜。

例一：

买方说："你们在法国还有其他的客户吗？"

卖方说："我公司已经进驻法国五大销售渠道，并占有法国市场12%的份额。"

买方不知道接下来应该说什么了。

卖方的回答是否已经无懈可击，完全占据上风了呢？其实，可以讨论的线索很多。例如，12%的市场份额，是处于上升的状态还是处于下降的状态？卖方的竞争对手实力如何，市场份额是多少？卖方的市场份额不少，有没有出现过负面的报道？……如果买方准备不充分，那么可以先把该议题记下来，休会时做进一步的研究，下次谈判时可作为一个新的议题/筹码。

例二：

对方要求赠送20件包装（预料之外的要求），我方马上回应不行。这样是不对的，对于意料之外的问题的回答，应经过大家的讨论。

正确的做法：对于预料之外的问题，若需要小组内部讨论，应示意对方，让我们再讨论讨论；若无讨论必要，至少小组内部有个示意可否的过程。

例三：

对方以我方股价下跌为由质疑你方的经营状况，马上就回应经营状况没有问题。正确的做法：首先，查询我方实时股价和大盘的走势；其次，查询对方实时股价和大盘的走势；最后，翻阅准备资料中双方近期的大事要闻。

## 二、发言没有步骤和重点

发言没有深度，语言没有层次。正确的做法是先在草稿上列出提纲或关键词，如第一、第二、第三分别说了什么。

举例：对方以2008年我方公司贿赂事件对我方进行施压，对我方公司的信誉表示质疑。我方可以从以下角度进行回应：1）对该事件表示遗憾；2）表示该事件属于个案，公司公正廉明的经营理念不会改变；3）希望对方能积极审视双方多年来的良好合作以及你方的优秀表现，给予客观评价。

## 三、发言没有针对性

选择性回避视听，发言没有针对性，回答和对方抛过来的话题对应不上。

举例：买方说："你方价格过高，这个价格我们卖不出去，是要亏损的。"卖方回答："我们的原材料来自国内A级产区，产品通过中国和欧洲的相关认证，在质量上是绝对有优势的。"

卖方的回答有什么问题呢？这个回答的内容和买方的内容是对应不上的，没有围绕对方关心的主题——价格来进行。事实上，卖方的回答就像一个模板一样，跟对方说什么没有关系，只考虑自己想表达的内容。若没有"听"到对方的发言，回答不能答疑解惑，则起不到效果。在谈判中，立场坚定固然重要，但是思维的活跃，也是双方沟通、理解、互利共赢、加深和扩展合作的关键。

## 四、团队没有主心骨、发言无序

谈判时出现几个人抢话、插话的情况，谈判主题逐渐偏离主线，或者话题频繁更换，每个话题都没有深入探讨甚至没有讨论出结果就被更换。正确的做法：每个小组有一个核心人物作为谈判的总指挥，发言之前小组内部先有个示意，有一个眼神交流、发言、补充的过程，每个话题至少有阶段性的结果才能更换。这个问题可以随着谈判经验的增加或老师预警或纠正之后得到改善。必要时可更换主谈人。

## 五、准备不充分

在谈判过程中只有态度，没有数据和事实支撑。表现为：不管怎样，就是不让步、不同意，问原因又说不出个所以然，让对方感觉谈判人员不专业、无法进行深入讨论。

## 六、谈判态度不正确

有的小组态度过于强势，双方在某个问题上僵持太久，导致谈判过于艰难。为预防此种情况的发生，应设定一个原则：双方要遵守合作的精神。可以根据需要适当给对方压力，但不可过之。在对话时不可无缘无故地一直说不，若对对方给出的理由无法反驳，则应答应或休会，讨论应对措施。

有的小组过于好说话，什么条件都答应，以致谈判很快就结束了，双方都没有发挥的空间。若要预防此种情况的发生，则需要双方对合同的条款、行业惯例、国际市场等内容做更深入的研究。毕竟在国际贸易中，观念冲突、市场信息失衡、商业惯例迥异、公司和个人需求差异、个性冲突等问题是很常见的。

---

**沟通发言技巧**

当对方说："你的样品很好，但是尺寸有点太大了。"你该如何回答？

首先，感谢对方对我方样品的肯定。

其次，询问对方合适的尺寸是什么？商品的用途是什么？再据以推荐合适尺寸的样品或者与对方商量尺寸的调整方案。

注意：很多同学会直接说："我们的质量很好。""我们的用户评价很好。"这样的回答不是最优的，因为我们还不知道客户的真正需求是什么，可能表述的不是对方关注的内容，甚至造成"各说各的"局面，沟通效率很低。

---

# 附录一  国际贸易销售合同范例

## 《国际贸易销售合同》

合同编号：　　　　　　　　　　　　签订日期：

1. 买方：

地址：

电话：　　　　　　　　　　　　　传真：

2. 卖方：

地址：

电话：　　　　　　　　　　　　　传真：

经买卖双方确认根据下列条款订立本合同：

| 3. 商品名称及规格 | 4. 数量 | 5. 单价 | 6. 总金额 |
|---|---|---|---|
|  |  |  |  |
|  |  |  |  |
| 总计 |  |  |  |

7. 总值（大写）：

8. 允许溢短装＿＿＿＿＿＿＿％。

9. 成交价格术语：

□FOB　　　　□CFR　　　　□CIF　　　　□DDU

10. 包装：

11. 运输唛头：

12. 运输起讫：由＿＿＿＿＿＿＿＿（装运港）到＿＿＿＿＿＿＿＿（目的港）。

13. 转运：□允许；□不允许

14. 分批：□允许；□不允许分批

15. 运输时间：

16. 保险：由＿＿＿＿方按照发票金额的＿＿＿＿＿％投保＿＿＿＿＿险，加保＿＿＿＿＿险，从_____到_____。

17. 付款条件：

（1）买方应不迟于_____年____月____日前将100%货款用即期汇票/电汇支付给卖方。

（2）买方应于_____年____月____日前通过_____银行开立以卖方为受益人的_____天不可撤销信用证，有效期至装运后_____天在_____议付，并注明合同号。

（3）付款交单：买方应凭卖方开立给买方的_____期跟单汇票付款，付款时交单。

（4）承兑交单：买方应凭卖方开立给买方的_____期跟单汇票付款，承兑时交单。

18. 单据：卖方应将下列单据提交银行议付/托收。

（1）整套正本清洁提单。

（2）商业发票一式_____份。

（3）装箱单或重量①单一式_____份。

（4）由_____签发的质量与数量证明书一式_____份。

（5）保险单一式_____份。

（6）由_____签发的原产地证明一式_____份。

19. 装运通知：一旦装运完毕，卖方应立即电告买方合同号、品名、已装载数量、发票总金额、毛重、运输工具名称及启运日期等。

20. 检验与索赔：

（1）卖方在发货前由_____检验机构对货物的品质、规格和数量进行检验，并出具检验证明。

（2）货物到达口岸后，买方可委托当地的商品检验机构对货物进行复验。如果发现货物有损坏、残缺或规格、数量与合同规定不符，买方须于货物到达目的口岸的_____天内凭_____检验机构出具的检验证明向卖方索赔。

（3）如买方提出索赔，凡属品质异议须于货物到达目的口岸之日起_____天内提出；凡属于数量异议须于货物到达目的口岸之日起_____天内提出。对所装货物所提的任何异议应由保险公司、运输公司或邮递机构负责的，卖方不负任何责任。

21. 不可抗力：如因人力不可抗拒的原因造成本合同全部或部分不能履约，卖方概不负责，但卖方应将上述发生的情况及时通知买方。

22. 争议的解决方式：任何因本合同而发生或与本合同有关的争议，应提交中国国际经济贸易仲裁委员会，按该会的规则进行仲裁。仲裁裁决是终局的，对双方均有约束力。

23. 法律适用：本合同的签订地，或发生争议时货物所在地在中华人民共和国境内或被诉人为中国法人的，适用中华人民共和国法律。除此之外，适用《联合国国际货物销售合同公约》。

本合同使用的 FOB，CFR，CIF，DDU 术语遵守国际商会《2020 年国际贸易术语通则》的规定。

_____

① 为体现与国际贸易工作的一致性，本书遵从行业习惯的称谓，书中的重量、净重、毛重实际指质量、净质量、毛质量。

24. 文字：本合同中文、英文两种文字具有同等法律效力，在文字解释上，若有异议，以中文解释为准。

25. 附加条款：_____（本合同上述条款与本附加条款有抵触时，以本附加条款为准）。

26. 本合同共_____份，自双方代表签字/盖章之日起生效。

买方代表人 　　　　　　　　　　　　　　　　卖方代表人

签字（盖章）　　　　　　　　　　　　　　　　签字（盖章）
　年　月　日　　　　　　　　　　　　　　　　　年　月　　日

# 附录二　铁矿石购销合同范例

## 《铁矿石购销合同》

本合同签订于＿＿＿＿＿＿＿

买方：

地址：

电话：

卖方：

地址：

电话：

### 1. 产品供应、数量、交付

1.1　卖方责任和产品从卖方矿场的交付。

1.1.1　依照本合同的条款和条件，在合同年内，卖方应进行销售和交付，而买方应进行购买、领货、支付产品年度数量的金额，这个数量是双方在＿＿＿所同意的。卖方按照＿＿＿所签订的数量将货物送至买方，计算方式是离岸价（FOB），交到约定的装货港，并且依照装运条件交货。从卖方交给买方后，产品所有权和所有相关的风险，都应依照本合同的内容，从卖方转给买方。

1.1.2　按本合同所卖和所买的产品，都应是卖方在卖方矿场所开采和生产的。

1.1.3　在不损害本合同提出的年度数量制定程序下，双方同意并有认知，依照本合同的产品年度买卖数量，按照下列条款，在有限追索性基础下，将有足够的数量可使卖方取得对矿场所给的融资。

1.2　交货失败的通知。

1.2.1　如果卖方在任何时候知道无法按照合同要求生产，在约定好的日期交给买方时，卖方应立即以正式书面的方式通知买方，说明无法依照相关订单交货的重点原因。

1.2.2　在卖方无法依照本合同内容的产品数量交货时，卖方须支付买方所有的成本、损害、费用、责任、损失、处罚或罚款，包括但不限于空载运费和/或支船运费，这些都是买方因为卖方没有交货所产生的费用。

1.2.3　相对地，如果买方无法依照合同的计划，指定和购买产品，买方应向卖方支付所有的成本、损害、费用、责任、损失、处罚或罚款，包括但不限于销售上的损失，这些是卖方因为买方没有购买所产生的费用。

### 2. 不可撤销信用证

2.1　买方对付款责任的保证。

2.1.1　为了确保下述付款责任的准时完成，买方同意在合同年每一季开始的＿＿＿天内，开给卖方受益者的信用证是自由议付、不可转让、不可撤销、见票即付的。

2.1.2　卖方在合同年每一季的 LC 准时发出后，其交货的责任成立。

2.1.3　每一次的 LC 应持续有效，一直到卖方对于合同年相对季的责任完成，或者直到卖方和买方同意并且书面制定的其他条款的期限结束。上述情况以先发生者为准。

2.1.4 合同年每一季和各季的最后一个营业日，买方应提供给卖方可接受的证据，证明 LC 仍然有效。

2.1.5 在可接受的银行不再是可接受的情况下，卖方应书面通知买方，以便买方能安排一家卖方能接受的国际一级金融机构来开具 LC。

### 3. 样品和分析

3.1 装货港的抽样和分析。

3.1.1 对于每一次产品的装运，应进行一次代表性的抽样和物理化学方面的分析，应该由卖方自己花费在装货港完成。

3.1.2 买方有权利自己花费指派一位代表（经过卖方批准，而且这样的批准不应无理地扣留或延迟），参与抽验和物理化学分析的过程。

3.1.3 卖方所执行的物理化学分析，应由卖方自己花费，并且以装货港所发的相关分析证明为基准。卖方应在货船从装货港出港后的_____个营业日内，将这份分析证明通过 E-mail 寄给买方。

3.1.4 卖方应在装货港抽取一份现货样品，并保留至少_____日，从相关货品从装货港出港日开始算起，以便有需要进行化学或物理分析的仲裁使用。

3.1.5 装货港的抽样程序应是买方和卖方同意的。装货港的物理化学分析应根据可适用的 ISO 标准程序进行。

3.2 卸货港的抽样和分析。

3.2.1 对于每一次产品的装运，应进行一次代表性的抽样和物理化学方面的分析，应该由买方自己花费在卸货港完成。

3.2.2 卖方有权利自己花费指派一位代表（经过买方批准，而且这样的批准不应无理地扣留或延迟），参与抽验和物理化学分析的过程。

3.2.3 买方所执行的物理化学分析，应由买方自己花费，并且以卸货港所发的相关分析证明为基准。买方应在货船完整卸货后的_____个营业日内，将这份分析证明通过 E-mail 寄给卖方。

3.2.4 买方应在装货港抽取一份现货样品，并保留至少_____日，从相关货品在卸货港完整卸货日开始算起，以便有需要进行物理化学分析的仲裁使用。

3.2.5 卸货港的抽样程序应是买方和卖方同意的。卸货港的物理化学分析应根据可适用的 ISO 标准程序进行。

3.3 货物最终分析结果。

3.3.1 如果装货港所发的分析证明和卸货港所发的分析证明之间的差额等于或大于 1.0%（百分之一），双方保留用来进行仲裁分析的装封样品，应将其送至一位国际认定且买卖双方同意的仲裁分析员，对这些样品进行分析。

3.3.2 如果在卸货港没有进行分析，或者有进行，但是买方没有在完成卸货后的_____个营业日内，将检验分析证明寄出来，卖方在装货港所进行的分析将被认定为最终证明且对双方都有约束力。

### 4. 重量测定

4.1 装货港的重量测定。

4.1.1 对于每一次产品的装运,卖方应根据可适用的国际做法,由卖方自己花费,在装货港的相关船内,以水尺计重方式测定产品的数量。

4.1.2 买方有权利自己花费指派一位代表(经过卖方批准,而且这样的批准不应无理地扣留或延迟),参与重量测定的过程。

4.1.3 卖方在装货港所执行的重量测定,应由卖方自己花费,并且以提单和装货港所发重量证明为基准。卖方应在货船从装货港出港后的_____个营业日内,将这份重量证明通过 E-mail 寄给买方,并根据这份文件来开具相关发票。

4.2 卸货港的重量测定。

4.2.1 对于每一次产品的装运,买方应根据可适用的国际做法,在卸货港的相关船内,由买方自己花费以水尺计重方式测定产品的数量。

4.2.2 卖方有权利自己花费指派一位代表(经过买方批准,而且这样的批准不应无理地扣留或延迟),参与重量测定的过程。

4.2.3 买方应在货物完全卸货日开始计算,_____个营业日内,将卸货港所进行的重量测定的证明通过 E-mail 寄给买方。

4.3 货物最终重量。

4.3.1 最终且对双方都有约束力的重量应以下列方式决定。

(i)如果装货港和卸货港中货物的干重量之间的差额小于或等于 0.5%(百分之零点五),以装货港所发的重量证明所标示的重量为准。

(ii)如果装货港和卸货港中货物的干重量之间的差额大于 0.5%(百分之零点五),或等于 1.0%(百分之一),以装货港和卸货港各自发的重量证明上所标示的重量的算术平均值为准。

(iii)如果装货港和卸货港中货物的干重量之间的差额大于 1.0%(百分之一),双方应讨论和同意被认定为最终且对双方都有约束力的结果。

4.3.2 如果在卸货港没有进行水尺计重,或者有进行,但是买方没有在完成卸货后的_____个营业日内,将重量证明寄出来,卖方在装货港所进行的重量测定的结果将被认定为最终证明且对双方都有约束力。

**5. 价格及价格的调整**

5.1 双方同意以下价格并以美元支付所有数额:价格是每干吨(净重)FOB_____(港口),_____美元,价格根据以下项目进行调整并支付给卖方。

5.2 价格的调整。

价格应每_____(月/星期)调整一次。价格调整的依据是《金属导报》或《我的钢铁周报》的铁矿石价格指数或其他的主要国际价格标准。第一次价格调整将于_____(日期)进行。

5.2.1 铁。

如果铁含量每超过基本铁含量(62.5%)的 1%,则每吨价格应上调_____美元。如果铁含量每低于基本铁含量(62.5%)的 1%,则每吨价格应下调_____美元。

如果铁含量低于 60%,货物将被拒收,卖方有责任对买方进行补偿。

5.2.2 三氧化二铝。

三氧化二铝的含量应在 1.2% 以内,否则每超过 0.1%,则每干吨降价比例为每吨_____

美元。

5.2.3  二氧化硅。

二氧化硅的含量应在4.5%以内，否则每超过0.1%，则每干吨降价比例为每吨____美元。

5.2.4  磷。

磷的含量应在0.07%以内，否则每超过0.01%，每干吨降价比例为每吨_____美元。

5.2.5  硫。

硫的含量应在0.07%以内，否则每超过0.01%，每干吨降价比例为每吨_____美元。

5.2.6  锰。

锰的含量应在0.06%以内，否则每超过0.01%，每干吨降价比例为每吨_____美元。

5.2.7  水分。

如果水分超过8%，卖方将会按质量来补偿买方的损失，但雨季（每年的7—12月）来临时12%以内可接受。

5.2.8  物理特性。

(i) 当粒度超过50毫米的铁矿石超过部分5%时，每干吨降价比例为每吨_____美元。

(ii) 当粒度低于10毫米的铁矿石超过5%时，每干吨降价比例为每吨_____美元。

## 6. 整体

6.1  买方应依照本条内容所标的产品付给卖方，并且以美元来支付每次的货款。

6.2  发票。

6.2.1  卖方对每一次的货物应开一张发票（发票），金额是货物价值的100%FOBST。这份发票应基准于：

(i) 合同内容对货物所注明可适用的基准价格；

(ii) 该货物重量以装货港所发的重量证明中所测定的货物重量为依据。

6.2.2  一旦完成每次货物的装运，卖方应在相关船从装货港出港日算起，21（二十一）日内将下列文件，通过国际公认的快递公司寄给押汇银行和买方。

(i) 全套的清洁已装船"提单（B/L）原件的整份资料3（三）份，用来寄发给收货方。提单要求空白背书，通知方空白，并标示"按租船合同付运费"，同时标示出WMT的重量。

(ii) 基于100%FOBST金额和装货港货物重量所开的见运单，并标示船运公司的名称。

(iii) 根据合同内容所发的分析证明，标示出实际化学物理的分析结果，2（二）份正本，3（三）份影印本。

(iv) 装货港所发的重量证明，2（二）份正本，3（三）份影印本。

(v) 由＿＿＿＿＿＿＿＿＿＿＿＿＿＿＿＿＿＿或其他授权的地方当局签发的原产地证明，3（三）份影印本1（一）份正本和3（三）份影印本。

6.2.3  从B/L签发日算起的5（五）个营业日，卖方应通过E-mail提供给买方运货的细节。

6.2.4  从B/L签发日起的10（十）个营业日内，卖方应通过E-mail将本条内容所标示的文件原本细节和影印本提供给买方。

6.2.5  相关发票应该在上述内容所列的全部文件提交后开具，根据每次货品所应用的LC的金额来减除，才可付款。

6.3 多付和少付的款项。

除非双方有另外的协议，任何有关货物，以卸货港的重量和最终分析证明做基准来调整后，多付或少付的款项都应偿还，无论是卖方给买方，还是买方给卖方，应越快越好，而且应在收到多付或少付的证明文件后的_____个营业日内完成，并且不违反合同内容。

6.4 拖欠。

相关方如果在到期日没有收到付款，应在这之后可以的时间内尽快出具一份书面通知。无论这样的通知何时到达，拖欠的一方应知道未履约是从没有在应该付款的那日开始算起的。

6.5 有争议的付款。

6.5.1 如果任何一方对合同的任何一项付款有所争议，该方仍应准时付清没有争议的付款，并且尽快以合理切实的方式向另一方书面说明款项争议的细节。

6.5.2 任何争议后决定需付的款项，未付方应在_____个营业日内完成，并且包括依照合同所计算的利息和罚款。

6.6 付款货币。

合同下所有的款项，将以美元缴付，并且以电汇资金作为可立即存取的资金。

**7. 物权转让和风险**

在合同下所进行交货的产品的物权和风险，只要所有供应的产品符合合同，没有任何扣押、赊账、担保物权、累赘、任何一种不利的诉求，除非是买方造成的，都应在产品进入装货港的买方船舷后，从卖方转移到买方。

**8. 不可抗力**

如因天灾、内乱、战争、恐怖袭击，或动乱、骚乱、罢工、封锁、恶劣天气、火灾、爆炸、政府行为，或其他类似原因，且为任何一方无法控制的原因而使得该方无法执行或推迟执行本合同，任何一方都无须为此承担责任，但受影响的一方应及时向对方发出通知，并在其能力范围内尽可能合理地继续采取行动。

**9. 未履约和终止**

9.1 终止事件。

有履约方可以在下列未履约事件发生时，向未履约方提出书面通知，终止本合同。

（i）另一方对于本合同下须付、无争议的款项未付金额总额超过_____美元，而这样的未履约情况在履约方向未履约方提出书面通知后的_____日内，仍然没有改善。

（ii）另一方的解散或拍卖，除非作为企业重组部分地解散或拍卖，而且所形成的企业能承担起所有原先方的责任。

（iii）另一方的破产，或因破产造成的司法重组或司法以外的重组适用破产法。

（iv）另一方对于本合同有实质性违反任何一条责任（除了本条（i）点所叙述的情况），包括卖方没有依照该合同年的年度数量供应，和买方没有依照合同年的年度数量购买，除非这样的违反情况能在未履约方收到履约方所寄的有关违反情况和改善此情况的要求通知后_____日进行改善。

9.2 任何一方皆不得在没有原因的情况下终止本合同。

**10. 争端解决**

10.1 总则。

10.1.1 双方任何从本合同引起的事项或异议而造成的争端（简称争端）应由受侵害

的一方依照条款给予另一方通知（简称争端通知）。双方应在收到争端通知的30（三十）日内，尽力善意地达成对该事项或异议的谅解。

10.1.2 如果争端不能依照条款解决，双方应促使各方的执行经理举行会议以便解决争端。如果争端不能在收到争端通知后的45（四十五）日内解决，任何一方可以要求争端依照条款解决。

10.1.3 除了因争端因素而明显阻止合同履行的情况，双方应继续履行各自在合同中约定的责任，直到争端依照本合同明文解决为止。

10.2 仲裁。

双方无法按照上述程序解决的事项或异议，除了双方在本合同有改变的或有得到双方同意的，所有异议事项都应交由仲裁解决，依照有效的_____仲裁制度，取得决议和最终的解决，所有的仲裁程序应以_____进行，并在_____进行。

### 11. 准据法

11.1 本合同以_____的法律为准据法，如有任何冲突该法律无效。本合同内的贸易用词应受本合同所述产品装运当时，国际商会制定的《2020年国际贸易术语解释通则》和补充文规定的管辖，并按照该通则解释。

11.2 明确排除联合国制定的《国际货物销售合同公约》的应用。

### 12. 保密性

本合同的双方同意本合同的内容是保密的，除了为了执行本合同，无论是任何一方都不得透露本合同的内容给第三方，除非是一方的子公司和/或客户。例外的情形还包括在律师的意见中证明这样的披露是法律和一方所属的证券交易所要求的，或是得到另一方事前的书面允许的。

### 13. 其他规定

13.1 修改。

除了在本合同有特别明言的，本合同的任何规定都不得修改、更改、弃权、释放或终止，除非本合约双方皆书面签署了同意。对于本合同任何规定的违背不可被任何一方弃权或释放，除非有得到另一方的书面同意。

13.2 完整合同。

本合同构成双方的整体协议并取代先前双方对于本合同事项之间的书面或口头协议，排除任何本合同排除的法律规定。双方承认没有受任何本合同中不含有的声明、保证或承诺的影响而签署本合同。

13.3 税。

13.3.1 任何和所有产品或有关本合同在_____的税务、责任，都由卖方担负。

13.3.2 任何和所有产品或有关本合同在产地国以外的税务、责任，都由买方担负。

| 卖方 | 买方 |
| --- | --- |
| 签字（盖章） | 签字（盖章） |
| 日期 | 日期 |

# Chapter I

# Themes of Simulated Negotiation

The course of International Business Negotiation Training needs to be completed in specific situations, so relevant industries and commodities should be selected. This chapter selects familiar commodities and large-scale companies, which is conducive to data collection and role entering. The model negotiation topics include two export themes and two import themes, namely, the export negotiation of Chinese PVC cartoon three-dimensional dolls, the export negotiation of Chinese cotton garments, the import negotiation of frozen white shrimps from Ecuador and the import negotiation of iron ore from Australia. The training items in Chapters III, IV and V are supposed to be based on a negotiation themes in this chapter.

The corresponding international trade contract samples are attached at the end of the book. Students can discuss the details of cooperation according to the contents of the contract text. The international trade sales contract sample in appendix 1 is suitable for theme 1, theme 2 and theme 3, and the iron ore trade contract sample is aimed at theme 4.

Tips: After reading each negotiation topic, you should think about the following questions.

1. What are the advantages and disadvantages of your own party and the other party? Is this negotiation more beneficial to one party or the balance between supply and demand?

2. What should we focus on and prepare for this cooperation? What might the other party focus on and prepare?

3. What are the contents that can not be conceded and can be adjusted? What are the contents that the other party may not give in and can not adjust?

4. What are the possible difficulties in the negotiation? There should be difficulty ranking.

5. What is possible to reach a smooth agreement in the negotiation, and how should it be arranged and utilized?

6. What other services can be improved to attract the other party's attention as one of the magic weapons of transaction? For example, anti-counterfeiting scheme, etc.

## Theme 1   Chinese PVC Cartoon Three-dimensional Dolls Export

### Negotiation Target Products

See Table 1 – 1 negotiation target products.

**Table 1 – 1   Negotiation Target Products – PVC Cartoon Dolls**

| PVC Cartoon Dolls Sample | Product Information |
| --- | --- |
| | • Professional customized PVC cartoon three-dimensional dolls with drawings and samples<br>• Minimum Order Quantity：5,000<br>• Price：3.2 yuan/piece FOB Huangpu, the price changes with the quantity<br>• Product Name：PVC cartoon three-dimensional doll<br>• Material：PVC and other plastic materials<br>• Product Packaging：foam, color box, etc（can be customized according to customer requirements）<br>• Customized products need to provide 3D views or 3D files, or send samples for customization |

### Negotiation Scene Settings

### 1. Seller：Company A in Guangdong

Since its establishment in 2012, the company has a 3-storey plant with a total area of 8,000 square meters, which can successfully complete a series of production processes such as enamel, oil injection, pad printing and packaging. Among them, the pad printing department is equipped with 50 semi-automatic production equipments, with a daily workload of more than 800,000 times; Dust free workshop to realize food grade production; Automatic packaging equipments, with a daily packaging workload of more than 200 cubic meters; Assembly automation of glue dropping department, with a daily output of more than 2 tons; With complete qualifications, the company obtained foreign export trade qualification in 2013, and passed Disney certification, International Council of Toy Industry (ICTI) certification, NB-CU certification, Supplier Ethical Data Exchange (SEDEX) certification, British Retailer Consortium (BRC) certification and ISO 9001 certification.

In 2017, the company's annual output value exceeded 100 million yuan. The overseas markets include the Middle East, Northeast Asia, Southeast Asia, North America, South America and Europe. It mainly produces two categories of products, injection 3D dolls and 2D glue drops. The glue drops mainly include custom seals, custom refrigerator stickers, custom key chains, custom coasters, custom bottle openers, custom luggage tags, custom 2D suction cup key chains, custom pendants, custom pen caps, etc. The company currently cooperates with brands including Walt Disney, Universal, Coca Cola, Three Squirrels, Jumptrack, Huawei, Walmart, etc.

## 2. Buyer: Company D in France

This French supermarket chain group has dozens of small supermarkets with complete varieties. As the saying goes, "Although the sparrow is small, it has all kinds of internal organs." The commodities sold in the supermarket can meet the basic living needs of residents. It is densely distributed, mostly located in or close to urban residential areas. In addition to selling well-known brand goods, there are more than 100 independent brands in the supermarket. The customized products are used for the sales of independent brands.

## 3. Cooperation Background

It's the first time of cooperation for both sides. For Chinese side, this is a good opportunity for customized small commodities to enter France. For the French side, finding high-quality suppliers for long-term cooperation is also an important guarantee for the sustainable development of the company in the post epidemic era.

## 4. Negotiation Place: France

### Background of the Negotiation \\\

## 1. Global Economic Situation and International Trade Characteristics in the Post Epidemic Era

In the post epidemic era, based on the problems of vaccine rationing, virus mutation and epidemic counterattack, the economic recovery of many countries is characterized by uncertainty and increased differentiation, and the global trade volume is also shrinking. Coupled with the rise of unilateralism and protectionism in some countries in recent years, the uncertainty of global economic policies characterized by the imposition of tariffs has increased, and trade frictions among major economies have continued. The positive signal is that the first phase of the Sino-US economic and trade agreement was officially signed in January 2020, which injected stability factors into the Sino-US and the world economy, enhanced global market confidence, and international trade activities gradually developed in a normal and stable direction. In 2021, Global trade began to recover. Thanks to the leading economic recovery, China bears greater responsibility in the global industrial chain and supply chain. At the same time, China is also a major "buyer" of global goods. Data shows that in the first five months of 2021, China spent 6. 72 trillion yuan on imported goods, a year-on-year increase of 25. 9%. It can be said that China is becoming a "stabilizer" of global economy and trade.

Source: sohu. com—Marine Sentient Beings: Blocking, Transporting! Edited and adapted by the author.

First, China plays an irreplaceable role in the global supply chain. According to the 2020 China Manufacturing Power Development Index Report released by the Strategic Information Center of the Chinese Academy of Engineering at the end of 2020, China's manufacturing power development index is 110. 84, which ranks third among the world's major manufacturing countries. Quality bene-fitisthe major weakness of the manufacturing industry for a long time. However, there is still room for improvement in the national manufacturing pattern. As the world's largest manufacturing country for 11 consecutive years, China is located in the key position in the upstream of the global production division system, has the most complete manufacturing industry cluster in the world, and has the advantages of cost saving, innovation and deep division of labor in the field of foreign trade. Company A is located in Guangdong, China, ranked No. 1 of China's top ten manufacturing provinces. The added value of manufacturing industries and the number of enterprises above designated size in the province rank the first in China. It has formed seven industrial clusters with an output value of more than one trillion yuan, and its innovation level is in the forefront of the country. The "14th Five Year Plan" for the high-quality development of manufacturing industry in Guangdong province issued on August 9th, 2021 puts forward the development goals of Guangdong to further promote the high-quality development of manufacturing industry, reshape industrial advantages, and strive to build an advanced manufacturing base at the world advanced level and an important innovation gathering place of manufacturing industry in the world.

Source: *Southern Metropolis Daily*—Golden Dragon Fish Disclosed Half a Year's Performance, edited and adapted by the author.

Second, due to the impact of the epidemic, factories in many countries couldn't start normally, so their dependence on China's export materials had increased sharply. However, since 2021, the prices of various raw materials have increased one after another. As of August 2021, raw materials such as plastic, copper, aluminum, iron, glass, new alloy and stainless steel have increased by more than 30% respectively, and soybeans and its products have increased by more than 50%. In addition, the congestion of sea transportation is serious, which lead to "one box is difficult to find" situation, and the sea freight is also rising, which is almost synchronized with the rise of raw material prices, resulting in a significant increase in the cost of international trade.

---

**Why Can Manufacturing Clusters Save Costs, Facilitate Innovation and Deepen the Division of Labor?**

1) In the industrial cluster, there are a large number of upstream enterprises such as raw material suppliers and equipment suppliers, close to each other, which is convenient for the comprehensive comparison and selection of production enterprises, so that they can be put into production at lower costs and faster speeds.

2) In the industrial cluster, the middle and downstream production enterprises, such as users and dealers at home and abroad, also gather here. They provide manufacturers with all kinds of the latest market information, product information and technical information. Therefore, the product distribution center is also an information distribution center. Enterprises in the industrial cluster can quickly and comprehensively understand the information of similar products at home and abroad, so as to quickly adjust the product structure and improve the design and variety, and get the initiative and inspiration of innovation.

3) In the industrial cluster, as there are a large number of upstream, middle and downstream enterprises and information resources, each enterprise can outsource non-key businesses, such as the production, product design and sales of other parts, so as to focus all its funds and energy on the point with the most advantage and finally made it the cheapest and best in the country or even in the world, forming the core competitiveness of enterprises.

### The Famous Suez Canal Ship Jam

On March 23rd, 2021, a 200,000 ton Ever Given freighter (Changci) of Chinese Taiwan EverGreen Marine Crop. blocked the Suez Canal, one of the world's shipping throats, resulting in the total paralysis of two-way traffic and the interference on global shipping for 12 days. According to the estimation of Allianz Group, a German insurance giant, the canal blockage may cause a loss of $6 – 10 billion per week in Global trade due to factors such as the extension of cargo delivery. Fig. 1 – 1 shows that the small excavator is saving a large cargo ship.

**Fig. 1 – 1 The Small Excavator Is Saving a Large Cargo Ship**

The two main materials passing through the Suez Canal include end products and raw materials. Terminal products include clothes, shoes and hats, electronic products, food and other materials. Raw materials include oil, natural gas, pulp, copper, etc. Public information shows that the oil transported through the Suez Canal accounts for 30% of all marine oil, and the natural gas accounts for 8% of the world market. The blockage of the canal affects the supply of these products and leads to a rise in prices.

Fig. 1 – 2 Shows the changes of sea freight from Ningbo to Los Angeles, USA.

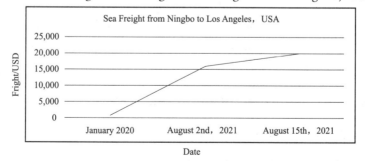

**Fig. 1 – 2 Changes in Sea Freight from Ningbo to Los Angeles**

Continued

When a 40-foot① container was transported from Ningbo, China to Los Angeles, USA, the price rised from about $1,000 in January 2020 to about $20,000 on August 21st, 2021. During the epidemic, orders from all over the world flew to China like snowflakes, but China's demand for goods from various countries was far from reaching the corresponding balance. Shipping companies are unwilling to return empty ships, which leads to a situation of "one box is hard to find" in Asia, especially in China.

According to the data released by CCTV 2 "World Finance and Economics", from June to July 2021, the freight rates from China to European destinations increased, see Table 1-2.

**Table 1-2　Freight Rates from China to European Destinations**

| Container Base Freight Rate | June 25th, 2021 | July 1st, 2021 |
| --- | --- | --- |
| Standard Container | $6,600 | $7,300 |
| Large-sized Container | $12,700 | $13,600 |

Source: WeChat Public "Yuanfang Qingmu" —Unremarkable Container, Became an Expensive Official Account of the United States, edited and adapted by the author.

## 2. Characteristics of China's Toy Export

China is the largest toy manufacturer and exporter in the world, and the EU is the main export market of China's toy products. According to the statistics of the General Administration of Customs of the People's Republic of China (GAC), from January to July in 2021, China's toy exports totaled $21,252.5 billion, a year-on-year increase of 51.5%. The export destinations were mainly traditional markets such as Europe and the United States, and the export volume to the EU was 53.14 billion yuan, a year-on-year increase of 75.9%. Exports to the United States amounted to 85.15 billion yuan, a year-on-year increase of 102.3%. In emerging markets, including Pakistan, Saudi Arabia, the United Arab Emirates, Kazakhstan, Sweden and South Africa, etc, China's toy export also achieved a rapid growth.

With the continuous innovation and development in function, design and various IP attributes, China's toy products will become more and more competitive in the international market in the future. In addition, with the development of new formats such as cross-border e-commerce, toy export has become more convenient, which will bring new changes to the overall development of the industry.

## 3. Characteristics and Trend of European Market

### 3.1　Characteristics of European Market

With a population of over 700 million, Europe is an important trading partner of China, com-

---

① 1 foot = 0.304,8m.

posed of 44 countries. Britain, Germany, France and other western European countries have a per capita GDP of more than $40,000, so they pay special attention to the quality of imported goods. As the per capita GDP of eastern European countries is relatively low, their consumers pay more attention to the cost performance of the products.

### 3.2  "Made in China" Is Recognized

"Made in China" was once synonymous with cheap and poor quality. However, after years of efforts, "made in China" has become a symbol of high quality and low price, and has a unique position in the world market. Under the epidemic, people's work, travel and living habits have changed, and some "made in China" products have become popular in Europe.

Take bicycles as an example. In order to avoid using public transport, bicycles with low price and stable quality have become the best choice for Europeans. However, the small number of bicycle production lines in Europe are backward in technology and can not meet the needs of the local market. Every year, bicycles made in China account for about 70% of the world's total amount. Chinese bicycles with exquisite appearance, excellent quality and reasonable price are highly praised by European consumers. It is reported that from May 2021, the export volume of Chinese bicycles and electric vehicles to Europe soared, and the "luxury" models at the price of ten thousand yuan were also sold out. Even if the manufacturers worked overtime, the production of many orders were still postponed to one month later at the end of 2021.

In terms of medical products, according to the analysis of German economic experts, since the outbreak of the epidemic, the medical products imported from China in Europe have increased by 50% year-on-year. In the first half of 2020, the expenditure on purchasing masks in Europe reached 14 billion euros, a year-on-year increase of 16.5 times. Chinese masks are favored by Europe. In addition to taking the lead in China's antiepidemic work, they also have obvious price advantages (the price of masks made in China is about 0.05 euros). The advanced protective equipment ventilators sold in China began to gradually replace the market of European local enterprises. China have also received orders from the largest ventilator manufacturer in Switzerland.

### 3.3  There Was a Great Demand for Toys During the Epidemic

During the epidemic period, due to the implementation of epidemic prevention and home isolation measures, people's home time increased, so the demand for toys increased greatly. According to the report released by NPD group, an American market research organization, its sales of 11 toy categories tracked in 12 global markets increased year-on-year in the first half of 2021. Among them, games and puzzles increased by 59%, outdoor and sports toys increased by 38%, and infant and preschool toys increased by 14%.

## 4. RAPEX Toy Products Notification

The safety of toy products is directly related to children's health and safety. Therefore, toys are listed as key regulatory products all over the world. In recent years, in the notification of the EU's "safety door" (RAPEX) system, toy products have ranked "first."

In the first half of 2021, RAPEX issued 1,116 notifications, accounting for 17% of the total notifications of toy products, and 147 notifications of toy products from China, accounting for 78.61% of the total notifications of toy products. But overall, the number of notified Chinese toy products decreased by 32.57% on a year-on-year basis.

In the first half of 2021, a total of 17 countries notified China's toy products. The top five notified products are: 51 cases in Poland, accounting for 34.69% of the notified toy products in China; 11 cases in Slovakia, accounting for 7.48%; 10 cases in France, accounting for 6.8%; 10 cases in Hungary, accounting for 6.8%; 10 cases in Sweden, accounting for 6.8%. The above five countries reported a total of 92 toy products from China, accounting for 62.59% of the total number of notified products from China.

According to the risk classification reported by RAPEX, in the first half of 2021, there were 168 risk types of toy products notified in China, and some products had multiple risks notified at the same time. Among them, most are asphyxiation risks (65 cases) and chemical risks (65 cases) caused by small parts falling off. Among the chemical risks, most are caused by pathalic acid esters in plastic toys and toys containing plastic that exceed the standard. In the first half of 2021, China's toy products were notified mainly because they did not comply with EU Toy Safety Directive [ (EU) 2021/1992], Safety of Toys—Part 1: Mechanical and Physical Properties (EN71-1: 2021), REACH Regulation (EC) No 1907/2006, Toxic Elements Test (EN71-3: 2021), Safety of Electric Toys (EN ICE 62115: 2020), etc.

Source: data released on RAPEX official website on July 1st, 2021, edited and adapted by the author.

# Theme 2   Chinese Cotton Garments Export

## Negotiation Target Products \\\\\

See Table 1−3, negotiation target products.

**Table 1−3   Negotiation Target Products – Cotton Garments**

| Cotton Garments Sample | Product Information |
| --- | --- |
| | • 10 customized cotton garments with OEM samples<br>• Minimum Order Quantity: 1,000 pieces<br>• Price: 18 yuan/piece FOB Hangzhou, the price changes with the quantity<br>• Product Name: cotton, short sleeve, female, V-neck T-shirt<br>• Material: 92% cotton + 8% spandex<br>• Pattern: hand nailed beads, 3D water print at the bottom<br>• Color: white, gray, black, rose, red<br>• Size: M L XL XXXL XXXL<br>• Package: waterproof plastic bag (can be customized according to customer requirements) |

### 1. Seller：Shanghai A Textile Import and Export Co. , Ltd

Founded in 1959, the company is one of the earliest state-owned professional foreign trade companies established after the founding of the People's Republic of China. It is also the core backbone enterprise of Dongfang International Group, the 2nd pilot unit of large-scale comprehensive trading company in the country. The company is mainly engaged in high-grade and multi-variety yarn, grey cloth and bleached cloth; Import and export trade of acrylic fiber, viscose fiber fabrics, woolen fabrics, blended fabrics and all kinds of clothing, finished textile products and other non-textile commodities. Since the 1980s, the company has exported more than 8 billion US dollars, always at the top of the list of the country's largest export enterprises. The company's trade involves a wide range. In addition to operating the import and export business of all kinds of textiles and their finished products and other non-textile commodities, undertaking compensation trade, incoming materials, incoming parts and incoming materials processing business, carrying out technical equipment introduction and technical exchange, and accepting domestic entrustment and import and export agency business, it also involves domestic trade, service trade, real estate management, advertising, industrial investment, scientific and technological development, consulting services and many other fields.

### 2. Buyer：Company B in America

This is an importer of American clothing brand. It has 36 branches in major cities in the United States. With stable operation and unique design style, it is welcomed by middle-aged and young American consumers. Company B plans to find a long-term cooperation company in China to produce 10 cotton garments by OEM, which has high requirements for quality. If the cooperation is successful, there will be a large number of orders every year.

### 3. Cooperation Background

The two parties met at Apparel Textile Sourcing (ATS) Miami in 2021, then they went to company B for a detailed discussion. That was the first time cooperation for both of them.

Exhibition Source：China Textile Import and Export Chamber of Commerce.

### Negotiation Place：America

### 1. Garment Exports Maintained a Good Growth Trend, but Profit Margins Were Squeezed

Textile industry is China's traditional pillar industry and has a great international competitiveness in the world. China is the largest textile and garment production and processing country in the world, and it is also a textile and garment export country. Affected by the Sino-US trade war, China's textile and garment export has withstood a certain pressure. However, due to the epidemic, orders from all over the world have returned, resulting in a new growth in China's textile and garment export.

According to the data released by the General Administration of Customs of the People's Republic of China on September 7th, 2021, counting in US dollars, from January to August 2021, the cumulative export of textiles and clothing was $1,984.68 billion, a year-on-year increase of 5.90%, breaking the record of clothing export in the same period. Among them, the export of clothing was $1,056.95 billion, an increase of 27.95%.

If the mask factor is excluded, the textile export achieved a positive growth of 49% in the first half of the year. In addition, the export of major commodities such as chemical fiber, yarn, fabric and textile machinery in the industrial chain all increased. The global economy continues to recover, driving the increase of external demand and boosting China's textile export.

However, Since 2021, the price of raw materials has risen sharply, the cost of shipping has risen significantly, and the operating pressure of many domestic export enterprises has increased considerably. The prices of textile and garment raw materials, such as cotton yarn and staple fiber, have risen all the way, and the price of spandex has increased several times over the beginning of the year, and the products are still in short supply. Cotton prices rose by more than 15% from late June to September 2021.

## 2. Maritime Logistics Is Seriously Blocked

Due to the hold-up of containers in Europe and in the United States, from June 2021, the situation of "one container is difficult to find" in Asia, resulting in a longer cycle of cargo transportation and a sharp increase in freight (see topic 1 for the increase in freight). "Now we can not order containers at all. It usually takes a month to 45 days for a batch of goods to be sent out. Some time ago, a batch of goods of our company were placed at the port for more than a month." Mr. Xie, an enterprise employee engaged in garments export, introduced in July 2021 that due to the lengthening of the shipping cycle, the enterprise's payment return cycle has also become longer. "In the past, payment could be received in more than 20 days, but now it takes at least more than 40 days. However, if the number of garment export is relatively small, you can choose international express and air freight logistics, and arrive in the United States in 5 – 7 days. For large quantities, you can choose a special sea line, with a cost of more than ten yuan per kg, and goods can arrive in the United States in 30 – 40 days."

## 3. China's Textile and Garment Export Structure

The top five markets of China's textile and garment export are the United States, ASEAN, EU, Japan and the Republic of Korea. In the first half of 2021, the five markets of textile and garment exports were $24.64 billion, $22.47 billion, $20.59 billion, $9.47 billion and $4.42 billion respectively, the year on year growth of 12.2%, 35.7%, −20.0%, −8.4% and 18.5% respectively.

The United States is the largest export market of textiles and clothing in China. In the first half of 2021, one fifth of the end products of the industrial chain were exported to the United States. According to the data of the Textile and Clothing Office of the US Department of Commerce, from January to May 2021, the United States imported $10.15 billion of China's textiles and clothing, a year-on-year increase of 32.3%, and continued to occupy the largest import source market of the United States.

## 4. Competition from Bangladesh

Bangladesh is the second largest garment exporter in the world after China. Since 2018, orders from American buyers in Bangladesh have continued to increase. According to the official data released by the Office of Textiles and Clothing of the US Department of Commerce, Bangladesh's garment export revenue to the US market was $5.4 billion in 2018, compared with $5.06 billion in 2017. "The Sino-US trade war has created a very good opportunity for Bangladesh's garment exporters. Bangladesh's garment export revenue will continue to soar in the trade conflict," Abdus Salam murshedy, former chairman of BEMEA, told the *New Nation Miscellany*. Abdus Salam pointed out that to American retailers, the tariff has led to an increase in the cost when they are purchasing and manufacturing goods from China. In order to control the cost, many American retailers are turning their acquisitions from China to manufacturing centers with lower cost such as Bangladesh.

## 5. The Development of High-end Areas of the Industrial Chain of China Is Still Relatively Weak

At present, the industrial chain of China's textile and garment industry is relatively complete and the comprehensive competitiveness is strong, but the advantages are still concentrated in the middle and low-end fields, while the development of high-end fields such as green environmental protection, high-end new materials and textile intelligent machinery is still relatively weak. In the early stage, China mainly relied on the labor-intensive model to develop the textile and garment industry. However, with the improvement of people's living standards, the prices of various production factors, including raw materials and human resources, have been increasing, and the price advantage has been declining, resulting in some orders beginning to be transferred to Southeast Asian countries and African countries with lower prices of production factors. Under the background of the complex international situation and the improvement of national scientific and technological level, the global textile industry value chain is showing the trend of reconstruction. Therefore, moving forward to the high-end field is an arduous task that must be completed for the sustainable development of China's textile and garment industry and further enhance its competitive advantage.

## 6. Six Basic Types of American Buyers

### 6.1   Department Store Buyer

Many American department stores will purchase all kinds of products by themselves, and different varieties will be purchased by different purchasing departments. Large chain department stores, such as Macy's, JCPenney, etc, almost have their own purchasing companies in various production cities. It is difficult for ordinary factories to access. They often choose their suppliers through large traders to form a purchasing system. Their purchase volume is large, the price requirements are stable, the change of products purchased every year will not be too large, and the quality requirements are very high. It is not easy to change suppliers. Most of them watch local exhibitions in the United States and will not go to China in person.

### 6.2　Mart

Mart buyers like Walmart, Kmart have a large amount of procurement. They also has their own procurement system (buying office) in the production market. Their procurement is highly sensitive to the market price, the requirements for product changes are also great, and the factory price is very low, but the quantity is very large. Factories with strong development force, low price and strong capital can focus on this type of customers. Small factories had better avoid doing business with them as the working capital of one single order will be too big to afford. In case the quality can not meet the inspection standard, it's hard to make a turn.

### 6.3　Brand Importer

Most brand importers purchase by themselves, such as Nike, Samsonite, etc. They will find factories with scale and good quality to place orders directly by OEM. Their profits are good, their quality requirements have their own standards, and their orders are stable. They will establish a long-term cooperative relationship with factories. At present, more and more importers in the world come to China's trade fairs to find manufacturers, which is worthy of the efforts of small and medium-sized factories. The business scale of importers in their country is a reference factor for their purchase quantity and payment terms before doing business. You can know their strengths through their website. Even small brands have the opportunity to be cultivated to big customers.

### 6.4　Wholesaler

Wholesale importers usually purchase specific products and have their own warehouse in the United States to sell their products through exhibitions. Price and product uniqueness are the focus of their attention. It is easy for such guests to compare prices because their competitors sell goods at the same exhibitions, so the price and product differences are very high. If the same products are sold at a lower price, they often run off the order because others have a lower price. The main procurement method is to go to China to see the exhibition and purchase. Many American citizens of China origin with abundant funds do wholesale business in the United States as wholesalers and do the purchase in China.

### 6.5　Trader

These customers may buy every kind of products, because they have different customers who will purchase different products, but the continuity of their orders is relatively unstable. The order volume is also relatively unstable. It's suitable for small factories.

### 6.6　Retailer

A few years ago, almost all American retailers purchased in the United States, but after the business entered the network era, more and more retailers make inquiries and purchase through the network. In fact, it is difficult to cultivate large customers, so it is more suitable for domestic whole-salers. It's only a waste of time for factories on quotation.

Source: FOB Foreign Trade Forum—North American Market, edited and adapted by the author.

---

### The Background of Sino-US Trade War and Its Impact on China's Textile and Garment Export Trade

**Beginning of Sino-US Trade War**

In January 2018, the Trump administration announced tariff increase of up to 30% and 50% for some imported solar photovoltaic products and large washing machines, respectively, and tariff exemptions for economies of other countries. According to the "301 investigation" report of the US trade memorandum with China, the relevant US departments will impose tariffs on about $60 billion of Chinese imports, which officially escalated the trade friction situation into a Sino-US trade war.

**The Impact of Sino-US Trade War on China's Textile and Garment Export Trade**

From September 1st, 2019, the United States began to impose a 25% tariff on the goods in the tariff list published in May, and these goods include textiles, clothing and related textile machinery with a large proportion of China's exports to the United States.

The United States is the largest exporter of China's textiles and garments. The previous textile and clothing tariffs were about 10% – 20%. The additional tariffs have brought great competitive pressure to China's export of labor-intensive products, and accelerated the transfer of orders to Vietnam, India, Pakistan and other low-end textile and garment markets.

---

## Theme 3   Importing Frozen Ecuadorian White Shrimps

### Negotiation Target Products

See Table 1 – 4 negotiation target products.

**Table 1 – 4   Negotiation Target Products – Ecuadorian White Shrimps**

| Ecuadorian White Shrimps Sample | Product Information |
|---|---|
| | Product 1: Ecuadorian white shrimps, 40 – 50 head/kg |
| | Product 2: Ecuadorian white shrimps, 50 – 60 head/kg |
| Planned import number: 200 tons for each type; refrigerated container; general packing: 10 boxes/carton, 2 kg/box | |

### 1. Seller：Ecuadorian E Company

A good-reputation Ecuadorian company, exporting about 9,000 tons of Ecuadorian white shrimps to China yearly, which has business relationship with many seafood import & export companies in China.

### 2. Buyer：Guangdong G Aquatic Co. Ltd

The company was established on March 1st, 2006, with registered capital of 1 million yuan. The company has its own complete seafood chain, and has been importing Ecuadorian white shrimps for many years.

### 3. Cooperation Background：Two Companies Cooperate for the First Time

### 4. Negotiation Place：Ecuador

### 1. Shrimp Market in China

#### 1.1　The Relationship Between China and Ecuador Is Very Deep

According to the data released by the General Administration of Customs of the People's Republic of China, since 2019, China has surpassed the United States to become the world's largest importer of shrimps (722,000 tons of shrimps were imported in 2019, a year-on-year increase of 179.8%). Among them, Ecuador, India, Thailand, Vietnam, Argentina, Saudi Arabia and Canada are the main countries for shrimp importing in China. Specifically, in 2019, China imported 322,000 tons of shrimps from Ecuador, a year-on-year increase of 321.1%, and the import volume was $1.86 billion, a year-on-year increase of 282.5%.

#### 1.2　Characteristics of Shrimp Market in China

China has a huge demand for seafoods, with imported shrimps accounting for more than 50% of the frozen shrimp market. Chinese consumers have increasingly strong demands for health and safety food materials, and the inspection and quarantine department has stepped up inspection. Some seafood enterprises have built a complete industrial chain including global procurement, processing, export and wholesale. The retail market is booming. E-commerce, supermarkets and franchise stores are the three mainstream channels.

### 2. Overview of Ecuadorian Shrimps Export Under the Epidemic Situation

#### 2.1　Quantity and Price in the Chinese Market

In China, in the first two months of 2020, the number of imported shrimps in China fell sharply due to the epidemic. Then it began to recover rapidly in March 2021. However, the import volume dropped sharply from July to August 2021, and increased again in November, with a year-on-year increase of more than 10%. However, the overall trend of fluctuation was stable. At the beginning of 2023, China adjusted the inspection regulations of imported food, and the quantity and price of imported shrimps hit a record high.

In terms of price, the price of white shrimps in Ecuador decreased by about 15% – 25% in 2020, lower than that of white shrimps in India, Indonesia and Thailand. In December 2022, the price of white shrimps in Ecuador and Vietnam rose significantly. At the beginning of 2023, Ecuadorian export of white shrimps continued to grow, and the amount exported to China increased by more than 40% year on year, but the export price fell by about 10%.

Source: Netease, edited and adapted by the author.

### 2. 2    Trends in the US Market

In June 2021, the average price of shrimps imported from Ecuador to the United States reached $ 7. 17/kg, an increase of 19% over June 2020. So far, Ecuador has been one of the lowest priced white shrimp suppliers in the United States, but now prices have increased even more.

Source: Aquatic Portal—Micro Water Network, edited and adapted by the author.

## 3. Import Formalities and Required Documents of Ecuadorian White Shrimp (Non-epidemic Situation)

### 3. 1    Preparation

It includes filing of consignees, filing of Chinese labels, the foreign producer registration at China's State Administration for Market Regulation.

### 3. 2    Documents Provided by Exporter

They are certificate of origin issued by the export country, aquatic sanitary certificate issued by the export country, foreign producer's registered certificate at China's State Administration for Market Regulation.

### 3. 3    Customs Supervision Conditions

For products of HS code 03061730, customs supervision conditions are A/B (Customs Clearance of Entry Commodities/Customs Clearance of Exit Commodities), namely, CIQ declaration is needed for Ecuadorian white shrimps import.

### 3. 4    Inspection and Quarantine Supervision Conditions

For products of HS code 03061730, the inspection and quarantine supervision conditions are P. R/Q. S.

P: entry animal and plant quarantine.

Q: exit animal and plant quarantine.

R: entry food hygiene supervision and inspection.

S: exit food hygiene supervision and inspection.

### 3.5    Import Declaration Elements

They are name of commodity, production or conservation method (frozen), state (shelled, un-shelled), Latin name, specification (for example, 41 – 50 pc/pound), and packing specification.

# Theme 4    Importing Australian Iron Ore

## Negotiation Target Products

See Table 1 – 5 negotiation target products.

**Table 1 – 5    Negotiation Target Products – Iron Ore**

| Iron Ore Sample | Product Information |
| --- | --- |
| | Product: <br> 62% iron ore fines (iron ore and concentrates, average particle size between 0. 8 – 6. 3 mm) <br> Planned Import Quantity: 6, 000 tons |

## Negotiation Scene Settings

### 1. Seller: H Company

Information from BHP Billiton Company can be used as a reference.

### 2. Buyer: A Steel Group

It is a leading enterprise of steel industry in China, the main purchaser of iron ore in China, as well as a major importer of iron ore in the world.

### 3. Cooperation: Cooperated for Years

Two companies have been cooperating smoothly for many years. But the negotiators of each side see each other for the first time.

### 4. Negotiation Place: Australia

## Background of the Negotiation

### 1. Import Volume and Source of Iron Ore in China

#### 1.1    China Is the World's Largest Buyer of Iron Ore

China has always been the world's largest buyer of iron ore. In 2020, China's new infrastructure investment increased, and the demand of iron ore downstream industry and steel industry increased, driving the overall demand of the industry. According to the calculation of Lange Iron and Steel Research Center, China imported 1. 17 billion tons of iron ore in 2020, and the external dependence of iron ore reached 82. 3%. With the continuous rise of domestic steel plant capacity, iron ore prices have reached new hights.

### 1. 2    High Import Concentration

China's iron ore imports come from more than 30 countries, but they are mainly concentrated in Australia and Brazil. Since 2014, about 80% of China's iron ore imports have been from Australia and Brazil, and the proportion has been increasing. In 2020, China's iron ore importd from Australia accounted for 66% of the total global imports and 21% from Brazil, forming a double oligopoly.

Source: Securities Star, edited and adapted by the author.

### 1. 3    China Has Taken Measures to Reduce the Import of Iron Ore

China has introduced relevant measures, including realizing the diversification of raw materials required by iron and steel enterprises, compressing crude steel production to reduce the use of iron ore, and further building overseas equity iron mines, so as to reduce the import demand for Australian iron ore. According to the data of the GAC, from January to July in 2021, China imported 650 million tons of iron ore, a year-on-year decrease of 1. 5%. Among them, 88. 506 million tons of iron ore were imported in July, with a year-on-year decrease of 21. 4%, which is the the 2nd consecutive month of decrease, and the month-on-month decrease of 1%, which is the 4th consecutive month of decrease.

Jiang Shengcai, secretary general of Metallurgical Mines' Association of China, pointed out that according to China's iron ore resource conditions, production conditions, technical level and other development conditions, China will have a certain capacity to increase production in the future. As long as there are policy support and measures in place, domestic mines are fully capable of ensuring 25%, 35% and 40% of the total demand in 2025, 2030 and 2035.

Source: Sohu—Today's Iron and Steel, edited and adapted by the author.

## 2. Australia's Economic Dependence on China

In terms of exports, China is Australia's largest export destination. Australia's 2017—2018 fiscal year data shows that Australia's exports to China account for 30. 6% of the total. It is mainly the demand for minerals, other resources and education services, with a total value of A $194. 6 billion.

In terms of imports, China is also Australia's largest source of import. In 2017—2018, Australia purchased goods and services from China, which is worthy of A $713 billion, including telecommunication equipments and parts, computers, furniture, mattresses, cushions, strollers, toys, game and sporting goods, accounting for 18% of its total import amount.

In 2020, Jeremy Thorpe, chief economist of Pricewaterhouse Coopers, published a paper on *China Matters*, explaining that if China's economy has a "hard landing", that is, GDP growth suddenly drops by 3%–5%, it may set off a "shock wave" in the South Pacific— Australia's GDP will directly lose A $ 140 billion (7%), and reduce 550,000 jobs.

Source: Sina Finance, edited and adapted by the author.

## 3. Import Formalities and Required Documents of Iron Ore

### 3.1  Documents Provided by Exporters

Exporters should provide certificate of origin, original bill of lading, contract, commercial invoice, packing list.

### 3.2  Customs Supervision Conditions

For products of HS code 26011120, customs supervision conditions are 7A.

7: automatic import licensing; A: customs clearance of entry commodities.

### 3.3  Inspection and Quarantine Supervision Conditions

For products of HS code 26011120, inspection and quarantine supervision is M, namely, iron ore import needs to do imported commodity inspection.

### 3.4  Import Declaration Elements

They are name of commodity, usage, processing method, appearance, ingredient content, average particle size, source (name of mining area), and date of signing the contract.

# Chapter II

# Personal Rapid Response Practice

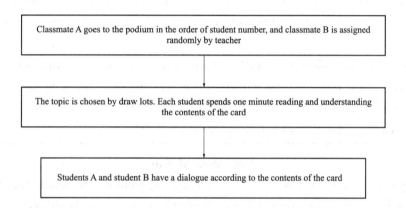

## Practice Mode

1) Practice combination. Each question needs to be completed by two students—student A and student B. Among them, classmate A is the trainee and completes the exercises by the order of student number. Student B comes to cooperate with student A, and can be assigned randomly by the teacher.

2) Title of the exercise item. It is shown in the form of card. The card is made by the teacher in advance. There are two cards for each question. Card A is for classmate A to read, and card B is for classmate B to read. Card A will explain the background of the topic, the reader's position and speech, and card B is student B's speech.

3) Draw lots. There are 25 questions in this chapter, one for each person. Each student will choose their own topic by drawing lots. Two students A and B stand on the podium together. After student A selects the topic by drawing lots, they take their own cards.

4) Reading cards. Two students each spend about a minute reading and understanding the contents of their cards.

5) Dialogue. Student A should first say the contents on card A, and then student B should say the contents of card B. The dialogue should be smooth and natural. After student B finishes reading, student A needs to answer student B in a short time. Student A can bring a piece of paper and a pen to clarify their ideas, and the thinking time shall not exceed one minute.

## Answer Steps and Examples

### 1. Answer Steps

During the negotiation, when the other party puts forward a topic unfavorable to us or a request that is difficult for us to accept, how can we answer to stabilize the scene, not only effectively fight back, but also have the other party listen to us, without annoying the other party? This textbook suggests the following steps: "Agree—But—Expression. "

Firstly, we agree with each other's views from a standpoint. After all, there are two parties to a business. It is understandable for the other party to understand the matter from their own point of view. Our understanding of each other is tantamount to giving each other a step down, allowing them to resonate and identify with us psychologically, and then they will accept our next views.

Secondly, show our difficulties or problems on the basis that the other party is being recognized by us in the first steps. Then, the other party would hear the facts stated by us, and even they would recognize and understand us psychologically.

Finally, put forward our views and refute the other party's views.

---

**A Moment of Fun**

Winston Churchill was a great man, but he had a big problem—he loved drinking. So he always quarreld with Mrs. Astor who advocated prohibition against alcoholic drinks.

One day, Mrs. Astor came up and said, "Winston, you're drunk again. It's annoying. "

Winston Churchill said, "Astor, you're right. I'm really drunk. But in the morning, I'll wake up, and you'll always be annoying. "

---

### 2. Using a Buffered Language

We all hope to give each other a comprehensive answer and take some time to clarify our ideas. Then, we can speak some buffer language and then each other has some time to think. For example, "I know what you mean, but I don't quite agree. Give me some time and I'll straighten my mind. " "I understand your difficulty, but I'm also very embarrassed. Let me think about what we should do. "

### 3. Answer Example

- **Contents of card A**

**Background**: You are the seller, in the negotiation room. Your company is not big, but the offer is very reasonable. You can not give in any more.

Speech: "For product T1258, our final quotation is $ 68, CIF Dubai. "

- **Contents of card B**

"Your quotation is reasonable. But please take a look at the quotation of company A. Although it is 1% higher than yours, it's a well-known large company after all. It's safe to cooperate. Let's consider it again. "

---

**A's Answer**

（Agree）I'm glad you received the quotation from company A. It seems that we have a lot of competition!

（But）However, 1% is not a small amount. Our business people are good at accurate calculation. If the service and quality provided are the same, I don't think you want to pay a higher price, do you? Besides, the large companies are always too proud. If you need to make any changes in the execution of orders, company A will not agree.

（Expression）However, we cherish every opportunity for cooperation. We are willing to pay more attention than company A, consider every detail, and let every customer enjoy better treatment. Although it is a small company, it has been operating in the industry for many years and will not damage our reputation for an order. You can rest assured.

---

Some answer ideas for reference for the above case are shown below.

1）Large companies also develop from small companies.

2）Small companies also care about reputation.

3）The state is supporting the development of small and medium-sized enterprises. For many countries, the contribution rate of small and medium-sized enterprises to GDP is very high, such as the United States, Italy.

4）Small companies cherish every opportunity for cooperation and can make flexible adjustments to future cooperation.

5）1% is not a small amount for the whole order.

6）Big brands will also have scandals.

7）Our raw material source is the same as company A/Our...is the same as company A.

Note: There can be many ideas to refute the other party, but 1 - 2 key ideas are enough in the answer. Do not stack all ideas.

## Questions of Scene Practice

**Question 1 Pick Up**

● **Contents on Card A**

**Background:** At 3pm, you pick up the other team at the airport. As one of the main negotiators, you hope to give them a good impression. You'd better talk about the cooperation or your company when answering them.

**Words:** "Nice to meet you, Mr. Abbott! How is the trip, is it comfortable? "

● **Contents on Card B**

**Words:** "Thank you for picking me up. The trip is good. It's the first time I come to China, China's development is much better than I expected. "

### Question 2 Exhibition Arrangement

- **Contents on Card A**

**Background:** You are an exhibitor in an exhibition. Your company just began the business in this industry 3 years ago. But you have experienced staff and hard-working people in the company. Please seize the opportunity and try your best to persuade him to go to your company.

**Words:** "I'm sure Mr. Smith is interested in our products, How about having a good talk in our company tomorrow?"

- **Contents on Card B**

**Words:** "Your products are really good. But your exhibition arrangement is too simple, inexperienced. Your company is really new, right?"

### Question 3 Psychological Tactics

- **Contents on Card A**

**Background:** You are the buyer, in the negotiation room. Your company is big with strong potentiality while the other side is a small company. Though your offer is not the lowest in the market, it's reasonable enough. The transaction price has an effect on your achievements and status in the company. You can not accept more concession on the price.

**Words:** "Through professional price calculation, our final offer is $25 per carton, CIF Long Beach."

- **Contents on Card B**

**Words:** "The price is not too high. But as you know, small companies don't make a lot of money. Big companies like you with advantages on costs and channels care more about service, isn't it? More reduction of 2% is not a big deal to you, but it will make us go whole hog."

### Question 4 Transportation Mode

- **Contents on Card A**

**Background:** You are the buyer, in the negotiation room. You hope partial shipment and transshipment is allowed as it's the peak time of sea transportation. Please stand your ground when answering.

**Words:** "We'd like to listen to your suggestion of shipment."

- **Contents on Card B**

**Words:** "Our requirement is partial shipment allowed, transshipment not allowed."

### Question 5 Make Mistakes

- **Contents on Card A**

**Background:** You are the seller, in the negotiation room. Your company has a good reputation and is financially strong. In the beginning period, the negotiation is going on in an orderly way. However, you find that the other side stop talking and whispers strangely with the material you provided. You need to be persuasive and rehabilitate your image on condition of admitting the mistake.

**Words** (**smile**): "What interesting thing are you talking about? Can we share it?"

- **Contents on Card B**

**Words**: "I am not picky. But you see, there is a wrong letter on your specification, we have to doubt about your executive capability. "

### Question 6: Price Decomposition

- **Contents on Card A**

**Background**: You are the seller, in the negotiation room. The products are western-style clothes. The suits in order A is new-style while order B is an old-style of last year; 200 pieces in an order. The two style of suits are slightly different in quality and material. Please take a firm stand after the other party's speech.

**Words**: "This is the quotation of order A and B, please take your time and read it. "

- **Contents on Card B**

**Words**: "Order A is 2,000 dollars more than order B! Are you kidding? Inflation doesn't cost so much!"

### Question 7 Cargo Damaged

- **Contents on Card A**

**Background**: You are the seller, in the negotiation room. The contract indicates that the products are transported in refrigerated container. What you leased is refrigerated container. The shipping company should be responsible for the lost. Please explain to the other party, calm them down and find solution with you together.

**Words**: "We are shocked about the loss of products. How much loss is it? How does it happen?"

- **Contents on Card B**

**Words** (**angrily**): "The contract indicates that the products are transported in refrigerated container. But 30% of the products go bad because of high temperature. Please give us an explanation. "

### Question 8: Price Reduction

- **Contents on Card A**

**Background**: You are the buyer, in the negotiation room. You ask for a further reduction of 1%, and insist your standpoint if they refuse it. But don't make them angry.

**Words**: "The price is still too high. 1% more reduction, we can sign the contract now!"

- **Contents on Card B**

**Words**: "The conditions are really unprecedented. If not for the economic recession of our country, we will never reduce the price to such a low level. Just to make money for porridge. If there is a further reduction, we'll go hungry. "

### Question 9 Be Nit-picking

- **Contents on Card A**

**Background**：You are the buyer, in the negotiation room. Please use critical strategy to be nit-picking on the washing machines in order to reduce price. This brand of washing machines are not well-known in China.

**Words**："What are the functions of this washing machine? Can you give an introduction?"

- **Contents on Card B**

**Words**："This brand of washing machines are able to warm the water and improve decontamination ability; it will clean the clothes through roller in a way of water conservation instead of using a lot of water; the dryer function is really useful to the humid weather in south China. "

### Question 10 Inadequate Preparation

- **Contents on Card A**

**Background**：In the negotiation room, you are a seller of refrigerator. You know that Korean brands have a great share in American home appliance market, but you are not sure about the specific proportion.

**Words**："This price is our bottom line. I don't know what are you hesitating about?"

- **Contents on Card B**

**Words**："In American home appliance market, Korean brands' influence is growing. In 2015, 77% of French door refrigerator priced over 4,000 dollars is Samsung brand. To promote refrigerator of Chinese brands, we need to spend more money and time. If you insist on this price, we can't accept it. "

### Question 11 Lose Temper

- **Contents on Card A**

**Background**：In the negotiation room. Your senior is a woman. The company's senior management is discussing about the deal intensely. You need to think how to handle the situation. You can be angry or patient.

**Words**："I'm sorry, the decision is still under discussion. Can you wait for a while?"

- **Contents on Card B**

**Words（angrily）**："Is your senior a woman? Dilly-dally, nothing new for a request for a whole day. We come from afar, not for Peking ducks!"

### Question 12 Payment Method

- **Contents on Card A**

**Background**：You are the buyer, in the negotiation room. Your company has good qualifications and you hope to conclude the deal with 30% T/T. Please try your best to persuade them.

**Words**："Our company always pays through 30% T/T. Last week, we just signed a big contract, more quantity than yours, with 30% T/T too. "

- **Contents on Card B**

**Words**: "I'm sorry, our company accept irrevocable L/C at sight only. "

## Question 13 Surprising Topics

- **Contents on Card A**

**Background**: You are the buyer, in the negotiation room. The other party reads your costing sheet carefully and proposes hard questions. Please stand your ground and refute their arguments.

**Words**: "It seems that Mr. Brady has been quiet for a while. Can you tell me what are you thinking about?"

- **Contents on Card B**

**Words**: "There is no problem on the costing sheet, but it's not the point. We had been to the Chinese supermarket and saw that the products were not placed on the main aisles but on a random corner. Let's make it frankly, why you don't earn money, is not because of the costing, but of your bad marketing policy. "

## Question 14 Mutual Liability

- **Contents on Card A**

**Background**: You are a buyer of red wines, in the negotiation room. You are talking about the issues of anti-counterfeit plan. Both parties agree to use Italian ID cork, but you hope that, the supplier is responsible for the purchase and the two parties share the added cost. They don't agree to share the added cost. Please try your best to persuade them.

**Words**: "Cork is used at the time of filing. I suggest that, you are responsible for the purchase. An Italian ID cork costs more than an ordinary cork does, we can share the added cost. How do you think about it?"

- **Contents on Card B**

**Words**: "There are so many fake wines in that area. That's the most in the world. Clients in other countries don't need Italian ID corks. We can do the purchase, but we won't share the added cost. "

## Question 15 Delivery Time

- **Contents on Card A**

**Background**: You are the seller, in the negotiation room. The earliest delivery that you can guarantee for the other party is next Wednesday. If they propose an earlier time, you have to refuse it and try your best to retain customers.

**Words**: "It's peak season of almonds at the moment and we get quite a lot of orders. But we can promise that the product will be delivered next week. "

- **Contents on Card B**

**Words**: "Our clients are badly in need of the almonds as the Christmas is coming. If they can't be delivered next Monday, we will lose a lot. "

### Question 16 Additional Requirements

- **Contents on Card A**

**Background**: You are the seller, in the negotiation room. The deal is expected any time, but the other party surprisingly proposed for 1 year shelf-life extension, which is not acceptable. Be care about your reply and don't restart the bargaining of key points as it will delay the conclusion for a long time.

**Words**: "The price is practical enough; we don't see a reason to refuse it!"

- **Contents on Card B**

**Words**: "OK, let's be straightforward! We accept the price and you extend the shelf-life of the products for 1 year."

### Question 17 Acceptance

- **Contents on Card A**

**Background**: You are the buyer, in the negotiation room. Your goal is a further reduction of 1.2%. If they agree with it, accept it. But don't do it too directly.

**Words**: "Another drop of 1.5%, let's close the deal!"

- **Contents on Card B**

**Words**: "It was hard this week and we all saw the sincerity and ability of each other. For our cheerful cooperation and long-term development, we make a last reduction of 1.2%. If you still can not accept it, we have to go home in vain."

### Question 18 Playing One off Against the Other

- **Contents on Card A**

**Background**: Guest court. Negotiation is at a key moment. As a general manager, you are invited by the chief negotiator of the other party for a drink at night. He might divide you and your teammate, don't be tricked and reply him skillfully.

**Words**: "Haven't been so relaxed for a long time. It's so tired these days. You are tired too, isn't it? Have a drink."

- **Contents on Card B**

**Words**: "Thanks. It's really relaxing here. You are a frank person, so there is something I need to remind you. Your chief negotiator, Mr. Chen, thinks differently with you. He cares about short-term benefit too much and makes the negotiation so hard. He's a talent, but his position and attitude makes me feel pessimistic about the negotiation."

### Question 19 Degree of Cooperation

- **Contents on Card A**

**Background**: You are a seller of refrigerator, in the negotiation room. You are talking about a consumer experience activity. You hope that the buyer could join the planning and holding of the activity, and bear 10% of the expenses. The other party would like to cooperate in the activity but refuse to bear expenses. Please stand your ground and emphasize that the sense of duty and result will be much different if they pay for the activity.

**Words**: "Consumer experience is the basis of R&D for next stage, as well as the basis of stabilizing and expanding markets for your company. We hope that you can cooperate with us in the planning and holding of the activity and bear 10% of the expenses."

- **Contents on Card B**

**Words**: "Consumer experience is really important. But it's important for you. We can give you a hand, but bear money? No."

### Question 20 Middle Man

- **Contents on Card A**

**Background**: You are the buyer, in the negotiation room. Guest court for you. Their general manager left for a business trip when the negotiation came to a deadlock. You communicate with their lawyer, enlighten him with feeling and emotional moving, hope he can help and break the ice.

**Words**: "I know that your general manager leaves on purpose. He wants to give us some pressure. We would like to yield, but the concession should be mutual."

- **Contents on Card B**

**Words**: "I understand your point, and I don't want to see the negotiation go in vain. But the general manager has his plans, I can do nothing."

### Question 21 Iron Ore

- **Contents on Card A**

**Background**: In the negotiation room. Your company is the buyer of iron ore, as well as a leading steel and iron enterprise in China, which means, your price of steel can affect the market price of the country. At present, China's steel market is having a great loss. Your company has to increase steel price twice to avoid large-scale close of small and medium steel enterprises. During the negotiation, the iron ore price is not reasonable enough. You have to keep your view points.

**Words**: "Mr. Smith also knows that win-win is the basis of long-term development. China's steel market is having a great loss, who can afford expensive raw materials?"

- **Contents on Card B**

**Word**: "You said you have a loss. But what I see is that China's steel price increased for two time this year. Why should we reduce raw material price?"

### Question 22 Negotiation in Other Places

- **Contents on card A**

**Background**: In the negotiation room. Your negotiation with the general manager assistant is coming to an end, but you feel that the price is too high. The night before signing the contract, you have a drink with the general manager. He's great-hearted. There might a room for maneuver. Please seize the opportunity and try to get a better price.

**Words**："We gave a lot of reasons but your assistant didn't understand it and wanted us to understand him in reverse. We feel that if we accept the price, we are forced to accept it."

- **Contents on Card B**

**Words（General Manager）**："Oh, my assistant is a little strong, I know. But, mutual understanding is the basis of closing the deal, isn't it?"

### Question 23 Ultimatum

- **Contents on Card A**

**Background**：You are the buyer, in the negotiation room. The negotiation is at your place and near its end. The price is close to your target level but the other party said they would leave. Insist on your target but don't offend the other party.

**Words**："We know your sincerity through the long-time negotiation. But the price is still too high."

- **Contents on Card B**

**Words**："The negotiation is still not successful even if we give so much concession. It's no sense of spending more time. We will go back by plane tomorrow. Contact me if there is something new."

### Question 24 Speed up the Process

- **Contents on Card A**

**Background**：In the negotiation room. Through watching their expressions, you know that the other party has accepted your condition. But they won't say it directly. You need to say something to promote transaction.

**Words**："I propose, for long-time cooperation, the technical fee is 85% of the market price in T country."

- **Contents on Card B**

**Words**："I see your sincerity and your change of conditions. But we have to make a study before giving you an answer."

### Question 25 Reject the Lower Price

- **Contents on Card A**

**Background**：You are the seller, in the negotiation room. At the end of the negotiation, the other party asks for a further reduction of 2%, which is not reasonable. Please stand your ground and refute it.

**Words**："We have a pleasant negotiation on the basis of mutual trust and appreciation. And you know that 2% reduction is not practical. Why do you make it?"

- **Contents on Card B**

**Words**："I believe that you know the consumption ability and potentials in US market. It's not about price but the long-term cooperation of the two companies. As a Chinese saying goes, 'Rangli' to do business, to show your sincerity, right? I hope I can see your sincerity."

## Answer Ideas for Reference

### Question 1 Pick Up

1. Affirm China's development.

2. Refer to the development of the industry or company.

3. Believe in the success of cooperation.

### Question 2 Exhibition Arrangement

1. Admit that the company has not been in this industry for a long time.

2. Emphasize that the company is guaranteed in product quality and service quality.

3. Exhibition products are only a part of the company's achievements.

4. For a more comprehensive understanding, please go to the company for details.

### Question 3 Psychological Tactics

1. Express understanding.

2. Large companies are also economical in operation.

3. If you don't reduce the price by 2%, won't you go all out?

### Question 4 Transportation Mode

1. Understand the other party's requirements.

2. However, under the current circumstances, such requirements are difficult to achieve.

3. The purpose of what we have done is ...

4. Ending (the ending is very important and related to the contents, attitude and position of the other party's answer in next round, so it should be positive and cooperative).

### Question 5 Make Mistakes

1. Admit the mistake, apologize and show the attitude of rectification.

2. I hope the other party will not question our overall execution ability with a mistake.

3. In the future ...

4. Ending (the ending is very important and related to the contents, attitude and position of the other party's answer in next round, so it should be positive and cooperative).

### Question 6 Price Decomposition

1. Understand the other party.

2. Generally, the new model is more expensive than the old one.

3. Split the price. $2,000 for 200 pieces, it's only $10 per piece.

4. The new model is the real demand of the market.

5. Ending (the ending is very important and related to the contents, attitude and position of the other party's answer in next round, so it should be positive and cooperative).

### Question 7 Cargo Damage

1. Please calm down.

2. We will never evade our responsibilities.

3. We did lease the correct container.

4. Conclusion (hope to have a good discussion and reduce the loss).

### Question 8 Price Reduction

1. Understand the other party's difficulties.

2. No deal is the real loss.

3. We all came to negotiate with a sincere attitude.

### Question 9 Be Nit-picking

1. Sounds good. But…

2. Heating…

3. Drying function…

4. Ending (the ending is very important and related to the contents, attitude and position of the other party's answer in next round, so it should be positive and cooperative).

### Question 10 Inadequate Preparation

1. Admit that Korean brand refrigerators are really popular.

2. But the market is not…

3. What you really need is…

4. Ending (the ending is very important and related to the contents, attitude and position of the other party's answer in next round, so it should be positive and cooperative).

### Question 11 Lose Temper

1. Apologize for the delay.

2. Explain the reason.

3. Protest against gender discrimination.

4. Invite them to eat Peking roast duck tonight.

### Question 12 Payment Method

1. Affirm the benefits of the letter of credit.

2. Our qualifications…

3. Limitation of the letter of credit: no error is allowed in the document, large amount of funds for the buyer, etc.

4. Ending (the ending is very important and related to the contents, attitude and position of the other party's answer in next round, so it should be positive and cooperative).

**Question 13 Surprising Topics**

1. Praise the other party's work.

2. The marketing work will be followed-up and improved.

3. Shelf placement is not a key factor in product promotion.

4. ...is the key to this cooperation.

5. Ending (the ending is very important and related to the contents, attitude and position of the other party's answer in next round, so it should be positive and cooperative).

**Question 14 Mutual Liability**

1. Admit that there are fake wines.

2. But the market is also the largest.

3. We have only...to maximize our interests.

**Question 15 Delivery Time**

1. We also hope to meet your requirements as much as possible.

2. Practical difficulties are...

3. The goods will be delivered next Wednesday.

4. Ending (the ending is very important and related to the contents, attitude and position of the other party's answer in next round, so it should be positive and cooperative).

**Question 16 Additional Requirements**

1. Express disappointment directly.

2. Both parties have paid a lot of time on...Why do they have other complications when closing a deal?

3. Ending (the ending is very important and related to the contents, attitude and position of the other party's answer in next round, so it should be positive and cooperative).

**Question 17 Acceptance**

1. Be silent for a while.

2. 1.2%... It's really difficult to make a job.

3. However, after talking for so long, we can also feel your sincerity.

4. Accept... And hope that in future cooperation...

**Question 18 Playing One off Against the Other**

1. Express surprise in the situation.

2. Say don't worry to the other party. Mr. Chen's point of view is consistent with the team.

3. I will also pay attention to the other party's ideas and will go back and unify my position with Mr. Chen again.

4. Ending (the ending is very important and related to the contents, attitude and position of the

other party's answer in next round, so it should be positive and cooperative).

### Question 19 Degree of Cooperation

1. Understand the other party's ideas.

2. Both parties are beneficiaries.

3. Activities involving interests will enable people to participate wholeheartedly.

4. Benefit maximization.

### Question 20 Middleman

1. Understand the other party's position.

2. Emphasize his role.

3. Both parties in the negotiation, including the general manager, hope that the negotiation will succeed.

### Question 21 Iron Ore

1. Recognize the price increase of steel.

2. Raising the price is still a loss.

3. Iron ore is purchased to support the future market, not today's market.

4. The other party is professional.

5. Ending (the ending is very important and related to the contents, attitude and position of the other party's answer in next round, so it should be positive and cooperative).

### Question 22 Negotiation in Other Places

1. It's good to understand each other.

2. But I prefer to get the opinion of the general manager himself.

3. The price is still high. I hope the general manager can decide.

4. A happy deal is the best deal.

### Question 23 Ultimatum

1. I didn't expect that the negotiation would reach this stage.

2. I still hope the other party can hold on for a while. After all…

3. Arrange tomorrow's activities, such as sightseeing, cocktail party and dinner, so that the other party can return home on another date.

### Question 24 Speed up the Process

1. OK! We believe there will be good news soon!

2. We also take this time to consider other terms of the contract.

3. Ending (the ending is very important and related to the contents, attitude and position of the other party's answer in next round, so it should be positive and cooperative).

## Question 25 Reject the Lower Price

1. It's too difficult for me.

2. Can't you see that we are sincere?

3. We attach great importance to US market.

4. Hope the other party can cherish the opportunity...

# Group-based Scene Practice

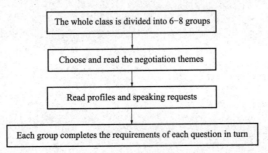

This session is group-based and six scenes are set according to the three stages of negotiation, opening, bargaining and closing. In each scene, each group searches for relevant materials according to the requirements of the topic, determines the contents of the speech, and makes corresponding scene demonstrations.

This chapter should be based on a negotiation topic in Chapter Ⅰ.

## Profiles

### 1. Grouping

The whole class is divided into 6 – 8 groups (about 5 – 7 people in each group is more appropriate, and the number of groups depends on the student number of the class), of which 3 – 4 groups act as the buyer and 3 – 4 groups act as the seller. Each scene is demonstrated in groups.

### 2. Division of the Group

A group is divided into the buyer subgroup and the seller subgroup according to the needs of the situation. During the demonstration, one student will speak, while other students will only cooperate and create a corresponding atmosphere without speaking. In actual negotiations, these scenes are also dominated by someone's speech rather than discussion. Therefore, in class practice, the speech of one student is also the main one, and other students can cooperate in actions and expressions to create a certain atmosphere.

In different scenes, each group's students take turns in speaking.

## 3. Roles of the Groups

Among the six scenes in this session, scene 3 "product introduction" is special, as all the six groups in the class complete the exercise as the seller. In the other five scenes, roles are conducted according to the original identities, ie, the buyer or the seller assigned.

**Request for Speech** \\\\\

> **Tips**
>
> The overall effect of scene practice consists of "speech contents" and "speaker's performance." The contents of the speech should take into account the integrity, focus, pertinence and other issues. The performance of the speaker is related to such factors as posture, aura, intonation, volume, speed and coherence.

## 1. Contents of Speech

1 ) The contents of the speech should be discussed and decided by the whole group, not by the students speaking alone.

2 ) The contents should be directly related to the subject. In the process of information searching, we will encounter a large number of complicated materials, some of which seem very tempting, but in fact, they are not directly related to this negotiation, nor are they the contents concerned by the other party. Therefore, it is necessary to filter the searched materials to ensure that they go straight to the subject and hit the point.

Taking the introduction of the beginning of the iron ore negotiation as an example. When searching the information on behalf of Chinese B Group, students will see a lot about the steel production scale and technical transformation of steel products, which B Group is proud of. Some groups will focus on this when making the speech. However, such an arrangement is not the most concern of the negotiating party, that is, the seller of iron ore. Although your party has a sense of momentum when talking, the other party sounds indifferent. Why not? Firstly, the target product of the negotiation is iron ore, not steel. Secondly, the advantages of B Group in the domestic market can not be directly equated to its position and voice in the international market. Then, what should be the focus and what to say in order to grasp each other's heart and arouse each other's interest? Please think and summarize by yourself.

3 ) The needs of both parties shall be taken into account, including objective and subjective needs, such as the situation of both parties, strengths comparison, concerns of the other party, the atmosphere we plan to create, establishing needs, arousing the interest of the other party in transactions, etc. For example, in terms of creating an atmosphere, we should think about what we need to show in this scene, whether we want to give each other pressure or temptation, what kind of atmosphere we want to create, relaxing or serious, frank or mysterious, etc.

## 2. Language Requirements

1）Speak without a manuscript. We should not speak in the form of "reading," we should be speaking in the way of "speaking." Such as "at this time, I think each of us should give the warmest applause..." and "about the main contents of today's meeting, I mainly talk about the following four points..."

2）Deliver a speech directly, don't say your train of thought. For example, at the beginning, he said on the stage: "Hello, everyone, I'm No. ×, × × × (name). Please allow me to introduce × × × (name)."

3）The speech should have depth, breadth and professionalism. Please compare the following sentences and see the differences between professional expression and non-professional expression. "Through large-scale production expansion, our dominant position in the market has been consolidated." vs. "The increase of litigation cases on such issues has not only led to the increase of legal costs, but also scattered the energy of our group management." vs. "Our quality is really very good."

4）Have a certain expression ability. If the same contents are demonstrated by different students, the results may be very different, which is related to expression ability. Expression ability is the comprehensive expression of manners, aura, intonation, volume, speed and coherence. Students' expression ability is related to their personal experience, speech ability and understanding of negotiation. It is hoped that each student's expression ability can be fully improved in this course.

## 3. Preparatory Work

The preparation for the questions is supposed to be made in class even if the class hours are only 24. If time is not enough for all the contents, the number of questions can be adjusted.

Effects of preparation in class are better as students can compare the performances of other teams and make the next answer better based on teacher's comments and guide. Students can adjust their direction of searching information, the way of selecting materials and the application of language and be able to improve their train of thoughts, professional levels and the expression ability, etc, gradually.

On the contrary, if the preparations are made before the class, neither the teacher's guide nor other teams' performances can be obtained as reference. Moreover, there is a strong possibility that the questions are not discussed and completed by the whole team due to the quick dividing of the work to each teammate. As a result, the answers' level is limited to the student's individual thought and level which is not the optimal result that we want.

### Questions of Scene Practice

It's allowed to make a 3 – 5 minutes' speech, 20 – 30 minutes to prepare for a speech.

## 1. Opening

Business negotiation opening refers to the period that both parties acquaint themselves with each other and issue statements respectively about the contents of negotiation before the formal discussion on the specific and substantive transaction items.

A favorable atmosphere should be created in the opening. It can be easy and harmonious, or pressured and serious according to the theme of the negotiation. Successful opening can lay a fine foundation for the negotiation in the future. The practice in the opening includes 3 questions, namely, opening introduction, opening statement and product introduction. In the practice, every team should determine the train of thought, specific contents and corresponding terms according to their position (buyer or seller), the background in the chapter 2 and online information. Some informal words can be used to adjust the mood.

### Scene 1: Opening Introduction

Please introduce your company and negotiators as a host.

**Contents:**

1) Your company: capability, products, operation scale, etc.

2) Members introduction: name, position, assigned tasks in the negotiation.

**Background:** It is the first time for the negotiators to deal with each other, but the two companies have cooperated for six years before, so you should be familiar with the strengths and cooperations of both companies. Please pay attention to the selection of speech contents.

### Matters needing attention:

1) As the host of the negotiation, you need to say some opening statements at the beginning and closing statements at the end of the negotiation.

2) For the members introduction, three members can be introduced, and students' real names can be used, but the corresponding positions and characteristics are prepared by students themselves.

3) It should be closely combined with the contents of the negotiation and not deviate from the subject. That is, every word you say is targeted and should be concerned about or related to each other. For example, when introducing the division of work for members, some groups mentioned that "classmate × × × (name) is responsible for recording the negotiation contents," which is the division within the group. There is no need to tell the other party, and it has no significance to promote the outcome of the negotiation.

4) The contents should be concise and you should pay attention to the overall fluency. The focus of the opening presentation is to create an atmosphere favorable to us, express sincerity, highlight our advantages, give the other party some pressure or temptation, and lay a good foundation for future transactions. For the company and product information, do not read down a long paragraph of the company's information, or make a detailed product introduction. In this way, not only the key advantages are not prominent, but also the on-site atmosphere is dull and distracted, which can not achieve the expected effect.

5) Attention should be paid to the application and fit of language. Please compare "Let me introduce the strengths of our company." "...to prove the strengths of our company." "The strengths of our company are as follows..." with "Let me briefly introduce our company below." Which sen-

tence sounds more natural and vivid?

　　6）It should reflect its own advantages. For example, it is inappropriate for the buyer to emphasize a significant increase in output when supply exceeds demand.

---

### Example of Opening Introduction

　　Hello, negotiators of H company! I am the sales representative × × × (name) from company D. On behalf of company D, I extend a sincere welcome to all negotiators. Next, allow me to introduce our negotiators. General sales manager × × × (name), chief financial officer × × × (name).

　　This is the sixth year of cooperation between our two companies, and the previous cooperation has achieved satisfactory results. I'm glad we're here again today. Our negotiators this time are all excellent staff selected by various departments of the company. I hope we can take over from the past and set a new course for the future, and cooperate successfully with your company again to create good results.

　　Your company is a first-class enterprise in the refrigerator manufacturing industry and has a good reputation. Your company's new refrigerator, with its function, appearance and the use of innovative technology, meets the needs of residents in the current market, and has great potential for sales development. With the strengths and experience of our company, if we can get the sole sales agency of your two new refrigerators in France, we will be able to successfully promote these products to the market.

　　In the next few days, I hope both of us can actively discuss this issue and gain something, so as to lay a solid foundation for your company's future sales in other regions and establish a higher brand influence in the French market in the future.

　　We have hired excellent Chinese translators × × × (name). If your company has any need in France, you can contact us, and we will try our best to provide you with our help. Once again, welcome all negotiators and wish us success in the next negotiations!

　　Source: Group 3, Class 1621, International Business Major, Guangxi University of Finance and Economics, Identity: refrigerator buyer

---

### Scene 2: Opening Statement

　　In the opening statement, both parties should reach a consensus on the negotiation procedures and related issues; both parties should indicate the intention and transaction conditions of one's own side respectively, understand the other party's scene and attitude through their statement and lay good foundation for substantive negotiation.

　　**Contents:** Standpoint, key problem, difficulties and hope should be contained in the opening statement.

　　**Notes:** Opening statement is a time to try to find out the real intention or situation of the other party. It should not involve specific transaction contents as it's used to point out a direction.

　　The following is an example of opening statement.

　　**Background:** India is developing rapidly towards the goal of super economy and has become the choice for many global computer and software companies to establish factories and R & D institutions. Bangalore, a city in southern India, has attracted world-class information and technology companies to invest. It has good transportation, rich talents and nice cultural environment, also known as "India's Silicon Valley."

However, in 2005, the city with a population of 6.8 million had only 2,700 rooms and suites, while Tokyo with a population of 12.7 million had 87,000 rooms in the same period. Millions of people come to Bangalore every year. The shortage of rooms makes Indian hotels the "most expensive hotel." The most advanced Leela Palace Hotel costs $390 – 450 per room for a night, and more than 90% of the rooms and suites are not left unused all year round.

If IBM employees of your company often go to India for 300 times a year, how can you persuade them to sign an annual accommodation agreement with your company?

Background Information Source: *International Business Negotiation*, Dou Ran, page 100.

---

### Opening Statement of Negotiation with Hotel Manager

I'm glad to have this meaningful negotiation with you. The goal of this negotiation is to sign an annual accommodation agreement for 300 people with your company to facilitate the travel of IBM employees.

As we all know, Bangalore, known as "India's Silicon Valley," is in a period of rapid development. As the world's largest information technology company, IBM is at the leading level in both software and hardware. Bangalore's development is inseparable from IBM's technologies and services.

Of course, we also know that your company's housing supply is very tight. But we can figure out the solution together. Further communication can be made in terms of cooperation methods. We hope your company will carefully discuss the details of cooperation with us from a win-win perspective. What do you think?

---

### Scene 3: Product Introduction

In this scene, 6 teams play the same role of the seller.

**Contents**: product' features, price trend, operation mode in the market, future market blueprint or other useful points.

**Matters needing attention**:

1) The contents should be screened. First, don't say too much about general characteristics of well-known products. After all, the other party is very familiar with and professional about the products. For example, in the iron ore negotiation, if elements and uses of iron ore are explained in detail, it will make the order reversed. Second, there should be several key contents, rather than equal distribution of space among all contents. Third, it can be introduced in combination with the contents of the quotation (some groups have prepared a virtual quotation by themselves, which makes introduction more realistic). Fourth, the comparative advantages with other suppliers should be clearly or implicitly indicated. Fifth, there should be some words to interact with each other, so that the other party has a sense of participation.

2）At the beginning of the introduction, the characteristics of the product should be classified. For example: "Next, I will introduce the characteristics of the product from three aspects, namely…" Each product has many characteristics. It is difficult to find the key point without classification.

3）It should be linked to the market and demand. Products with such characteristics are excellent, but they are not connected with the market and demand. The other party may not be aware of the value of this feature. For example, "In today's society, the demand for doubled-door refrigerators is increasing day by day, and our refrigerators adopt the principle of combining appearance and quality and unifying sales and services. Our designers emphasize the practical and versatile characteristics in style through the inspiration inspired by the visual creativity of French modern architectures, which is consistent with the recent aesthetic outlook of French consumers. It is consistent with the consumption guidelines. Of course, you can also compare the refrigerators of other suppliers to understand why you choose us."

4）We should accurately identify our position and job responsibilities. In international trade, the division of work for members and responsibilities between the buyer and the seller are very different from those in domestic trade. In the simulated environment, some students do not distinguish such differences at once, resulting in the dislocation of division on the contents of service and other issues in their speeches. If a refrigerator exporter says, "In terms of after-sale service, our company has realized door-to-door service within 2 hours." Can this be an advantage for expor-ters? In international trade, the buyer shall be responsible for the after-sale service of the refrigerators in the local market, or both parties shall be responsible through consultation. The seller can take this after-sale standard as a goal or suggestion to discuss cooperation with the other party.

5）The key advantages of the products should not be described directly, but should be emphasized or beautified in some corresponding languages, so as to arouse the other party's interest and leave a deep impression on the other party. Please compare "This refrigerator has passed × × certification, the only industry certification for refrigerators in the world." with "This refrigerator has passed × × certification, the only industry certification for refrigerators in the world. This is the result of our company's hard work and our pride."

## 2. Bargaining

Bargaining period begins from a quotation and ends with a conclusion or disagreement through courses of counter-offer, reoffer, recounter-offer, deadlock, and so on. Bargaining period is the core part of a business negotiation, as well as a process that both parties try their best to seek common points and reserving differences, to bargain and confirm the transaction conditions. At the same time, it's also a process of contest of preparation, bargaining strengths, negotiation experience and so on. Therefore, students need to consider the strategic element in the speech. The practice of bargaining includes 2 scenes, namely, counter-offer and deadlock handling.

### Scene 4: Counter-offer

Counter-offer is a demonstration made by offeree who proposes modification or change on the prior offer. It's an addition, limit or change on the conditions of the prior offer.

**Order of practice:** The buyer practice firstly. For example, the buyer propose a discount according to the exercise below, then the seller make a counter-offer based on the requirements.

**Points for reference:** Be nit-picking, demand from users, negative news, and so on. Other points are acceptable.

**Note:** The counter-offer should be a persuasive process instead of giving new conditions. The forms of speech are flexible. Students can be natural, happy, angry or modest in the speech.

### Exercise for buyer

**Background:** You get an offer from the seller: FOB $25/ton, CFR Qingdao $32/ton.

**Exercise:** Please propose a discount of 8%.

### Exercise for seller

**Background:** The buyer proposed a discount of 8 % on the offer (see the offer above).

**Exercise:** Tell them you can accept a discount of 2%.

### Scene 5: Deadlock Handling

A negotiation deadlock is a situation in which no progress can be made or no advancement is reached in the negotiation. A deadlock is often seen in the negotiation as some subjective factors like lack of communication, bias, and lack of understanding of the market are really difficult to avoid. Hence, sufficient communication is an important approach to break a deadlock.

### Exercise for buyer

**Background:** Your bottom line is CFR $58/ton while the other party insists CFR $60/ton. It's a hard time and you are really agitated.

**Exercise:** Please find an effective way to persuade the other party.

**Note:** No rules on the forms, methods and contents.

### Exercise for seller

**Background:** Your bottom line is CFR $62/ton while the other party insists CFR $60/ton. It's a hard time and you are really agitated.

**Exercise:** Please find an effective way to persuade the other party.

**Note:** No rules on the forms, methods and contents.

## 3. Concluding a Bargain

Concluding a bargain signifies a situation that all the parties reach an agreement on the bargaining problems. However, suspicions might be remained, for instance, is it the best condition that I can get, if there is still a better solution, whether the benefit of each party is balanced, how the other party regards the deal, and so on. Hence, we still need to do something and say something in order to remove doubts, satisfy the other party's psychological demand and establish a favorable basis for long-term cooperation which finally leads to an optimal benefit of the negotiation.

The practice of this part is about the closing statement.

**Scene 6：Closing Statement**

Both parties reach an agreement on price, transportation, delivery time, terms of payment, etc, through half month's negotiation.

**Contents：** Please sum up the efforts and achievements you make, and express good wishes.

**Notes：**

1）Specific contents about the deal are not needed as the closing statement is about the confirmation on both parties' efforts and future prospects.

2）Transaction is in sight. Something interesting or serious contents can be used to adjust atmosphere.

# Presentation of Negotiation Scheme （PPT）

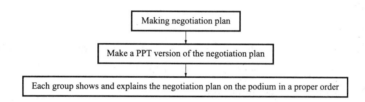

Continue the negotiation themes used in chapter I .

In the preparation phase of simulated negotiation, every team should search relative information on the market conditions, information of the other party of negotiation, the competitors of both parties and so on, make a negotiation plan PPT after summarization and hold an internal team mee-ting to display the negotiation plan to teammates.

The procedures of practice: Every team makes a negotiation plan PPT, and then sends a teammate to speak on the platform as the role of chief negotiator, to display the contents of the plan to the other teammates in the form of an internal group meeting. The time for preparation of PPT is 4 –5 class hours.

| **Tips** |
| --- |
| The display effect of the negotiation scheme consists of scheme contents, PPT production effect, and speaker's performance. |
| The contents of the plan is the basis for the group to prepare for further negotiations, which has a great impact on the outcome of the negotiations. The difficulty is higher than that in Chapter 2. |
| The production effect of PPT will be directly related to the speaker's play and the listener's understanding. |
| The performance of the speaker is the soul of the whole scheme. A good speaker can make a wonderful presentation of the general contents of production. Even facing the well-made contents, it is difficult for a speaker with the average ability to express the contents incisively and vividly. To improve all aspects of negotiation related abilities, more relevant exercises should be done. |

## Main Contents and Requirements

### 4.1　Main Contents of the Simulated Negotiation Plan

The following shows the main contents and requirements of the display.

1 ) Strengths comparison of both parties.

2 ) Advantages and disadvantages of both parties.

3 ) Product introduction.

4 ) Quotation ( including trade terms, combined with recent trend).

5 ) Market situation.

6 ) Work division for members.

7 ) Risk analysis.

8 ) Negotiation strategies, etc.

Except the tips in previous chapters, the following questions should be considered.

1 ) Which is more important, benefit or relationship?

2 ) Sequence and time of negotiation items.

3 ) Division of work for members includes assignment of negotiation items and use of strategies. Some special arrangements should be made as people have different personalities and styles like persuasive, impulsive, steady, friendly and serious. Some people are good at recording and concluding, some people are adept in easing a situation, some people can establish relationship quickly, and so on.

4 ) Some contents mentioned in the scheme are internal secrets, but it should be displayed in front of the whole class. Therefore, such contents like quotation can be adjusted slightly. Attention: contents and the sequence can be rearranged. Avoid listing the contents one-by-one. The contents should be key points and non-key points. Some contents can be deleted skillfully if necessary.

### 4.2　Length of Each Point

Display for every team is about 15 minutes. There should be 1 – 2 key points. Avoid averaging time on all the points.

### 4.3　Extracting Contents in the Plan

Every content in the plan should have been extracted and directly related to the negotiation. Avoid putting some information in the plan directly without a purpose. Please compare the statement "B Group has a wide market scope and abundant financial resources. In 2015, the Group reported a revenue of 1,637.9 billion yuan. " with "Payment capability is one of the points the seller cares most. B Group reported a revenue of 1,673.9 billion yuan …"

## Form of Speech

As a chief negotiator, you should know the contents of the plan well and display the plan confidently without looking at the draft. There is no rule on form and style of display. The display can be finished by a student alone from the beginning to the end or be finished by several people together with some interactive sessions.

The display should be made by "saying. " There should not be some informal words and personal opinions. For example, "We all know that, we can only lead the negotiation by understanding each other. " "I hope we won't take it lightly after we know our advantages," etc.

Rehearsals should be made 1 – 2 times before the display as few students are able to make an excellent speech just with terse words in PPT without preparation. Most of the students need a lot of practice to improve the level of making a statement. Moreover, rehearsals should cover all the pages in the PPT. If a student only rehearses the first few pages in the PPT, her/his performance on the latter pages is likely to be different.

## Language Requirements

1) The speaker is not supposed to speak dryly without paying attention to the listeners. She/he should consider if the listeners are really listening and understand what she/he said.

Please compare the following statements.

- "Push-back strategy. Let the other party speak first in order to understand their intentions, retreat in order to advance. According to their conditions ..." vs. "Now, let's talk about push-back strategy, whose key point is to let the other party speak first. The purpose of the strategy is to understand their intentions. This is the point, never speak first, and hold it on even if you're eager to say something. Understand? Can you do it?"

- "Two-man act strategy. Xiaolan Zhang plays the role of good guy while Jin Li acts as the bad guy. " vs. "Now let's discuss about two-man act strategy. The so-called two-man act, we all know that ... Xiaolan Zhang, your smile is so sweet, you should be the good guy for this task. Jin Li, you are majestic at ordinary times, you are the best choice of the bad guy...Has everybody cleared about your task?"

2) There should be some encouraging words. For example, "I suggest everyone here applauds loudly to thank our company leaders for their recognition of our abilities. "

## Risk Analysis

Risks include many aspects, such as the application of trade terms and payment methods, the other party's credit status, industry risks, macro environmental impact, legal risks, political risks, cultural risks, etc.

For example, in the selection of trade terms, we need to consider the exchange rate, transportation, delivery places, and so on. If RMB rises against US dollar, it will not have much impact on the foreign side in paying international freight. However, for the Chinese side, paying international freight after a period of time can save a sum of expenses. However, in general, we should try our best to arrange transportation by ourselves. For export contracts, if FOB terms are used for settlement, there is a great risk of collusion between importers and shipping companies and delivery of goods without the bill of lading. In addition, the use of foreign designated shipping companies, overseas freight forwarders or Non-Vessel Operating Common Carrier ( NVOCC ) is increasing day by day. However, many exporters do not understand the relevant national regulations and do not seriously review the qualifications of overseas freight forwarders or NVOCC. In fact, Ministry of Transport of the People's Republic of China (MOT) stipulates that the overseas freight forwarding bill of lading must be issued by a freight forwarding enterprise approved by China's relevant departments. The cargo owner can require the freight forwarding enterprise issuing the bill of lading to issue a letter of guarantee to release the goods against the original bill of lading at the port of destination. At the same time, it also stipulats that those who operate NVOCC business shall register the bill of lading at MOT which is governed by the State Council and pay a deposit. Overseas freight forwarders and NVOCC that do not meet the above conditions are illegal operators. For the importer, if the delivery is delayed due to the responsibility of the shipping company, the importer is not qualified to claim rights from the shipping company if using Group C trade terms.

Take another example of OEM production. Some exporters are hasty and only sign simple contracts when entrusted by well-known brands. In fact, such risks are great. Although well-known brands generally do not default, there will be business difficulties or even financial crisis, resulting in inability to pay. Such examples are real. In this case, what situation will exporters face? If a lawsuit of international trade is brought to the court, even if one party win the lawsuit, the other company will not be able to continue to perform the contract or bear the liability for compensation. If the goods are handled in a country or region where the other party has no registered trademark, first, there will be a great discount loss. Second, China has accessed to the Madrid Agreement Concerning International Registration of Marks, and exporters will face tort liabilities. So, how should risks be avoided? The best way is to sign a detailed contract. The licensing sales agreements of well-known foreign products often have dozens of pages, from equipments to processes, from standards to qualities, from accessories to the complete machine, from prices to markets, from service to maintenance, from delivery cycles to age limits of accessories supply, from the payment to delivery, from detections to acceptances, and trade-mark rights. Detailed and clear agreements should be made in advance.

Note: In this part, only one risk analysis is required for each scheme.

## PPT Design

### 1. Bad PPT Design

Fig. 4 – 1 shows the PPT of B Group's advantages analysis.

Problems in the PPT are shown below.

1) Too many words. It's difficult for the listeners to find the key points while the speaker is likely to "read" instead of "saying."

2) The contents are not closely related to the negotiation. For example, B Group's business scale and technological level has little to do with iron ore.

3) Lack depth in analysis. For example, there should be data with regard to the iron ore demand. When mentioning the other buyers, names and business volumes of recent years are essential.

## B Group's Advantages

- Scale advantages: B group is the biggest in China and the world's top 3 steel enterprise; it's business scope covers making and process of iron and steel and relative industries like electricity, coal, industrial gas production, wharf, storage, transportation and so on. As an important steel enterprise in Chi a, B group runs good with strong fund.
- Belongs to the world's biggest demand country for iron ore, has great influence on iron ore pricing.
- Advanced technology level and high purify on iron ore
- Have cooperation agreements with many iron ore enterprises in China and abroad, which of them are competitors of B Group

Fig. 4 – 1  B Group's Advantages Analysis

## 2. Good PPP Examples

Requirements of good design are concise, summarized, and highlighted. The following design is an example of good design. See the example of good PPT design in Fig. 4 – 2.

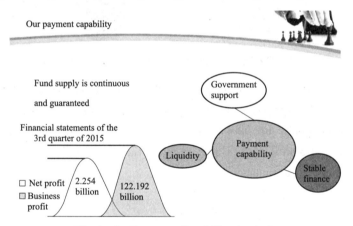

Fig. 4 – 2  Payment Capability Analysis

## Examples of Negotiation Schemes

Fig. 4 – 3 shows our strengths analysis.

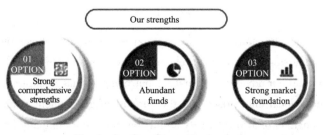

Fig. 4 – 3  Our Strengths Analysis

In terms of strength, our comprehensive strengths are relatively strong. As one of the high-quality leaders in the French electrical industry, we have abundant funds and have established a relatively solid market foundation in many years of cooperation. After understanding our strengths, we hope you can have a more confident sense of dependence in the process of negotiation. After clarifying our strengths and interests, let's take a look at the negotiating partner, namely China H group, and their interests and strengths.

Fig. 4 – 4 shows the other partys interests and strengths analysis.

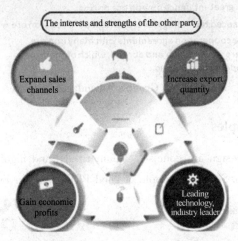

Fig. 4 – 4　The Other Party's Interests and Strengths Analysis

If China H group can successfully cooperate with us this time, it can also broaden its sales channels, improve its own export volume and obtain economic profits. In addition, the strengths of the other party are also strong, because it has high-quality integrated marketing channels and very good technologies. It is also an industry leader in the refrigerator brand market.

Picture 4 – 5 shows our advantages.

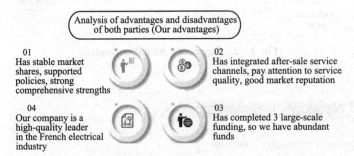

Fig. 4 – 5　Our Advantages Analysis

The following is about the analysis of the advantages and disadvantages of both parties.

On the basis of understanding our comprehensive interests and strengths, let's clarify our position in this negotiation and our advantages. Firstly, our company has a stable market share. Based on the rapid development of local enterprises and the support of government policies, we can say that our comprehensive strengths are also relatively strong. Second, our company has integrated after-sale service channels. Our company has been listed, pays attention to service quality, and the market reputation is relatively good. Third, in the past few years, our company has completed three large-scale financing, and the funds is relatively abundant. Fourth, there is no doubt that our company is a high-quality leader in the French electrical industry.

Based on the above points, I believe that the advantages of our company are very clear.

Of course, we say that there are both advantages and disadvantages. People are not sages and can not be perfect. After recognizing the advantages, we should also be aware of our disadvantages and shortcomings.

Fig. 4 – 6 shows our disadvantages analysis.

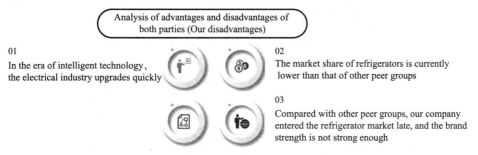

**Fig. 4 – 6　Our Disadvantages Analysis**

First of all, in the era of intelligent technology, with the rapid renewal of technologies, the renewal of electrical appliances is very fast. For the new refrigerator products exported, our investment in import sales is relatively large and the risk is high.

In addition, we are not in a leading position in the refrigerator market. The market share is lower than that of other peer groups. It can be said that we are still at the middle and lower level.

Finally, compared with other groups at the same level, our company entered the market a little later, and the strength of the brand is still not strong enough.

After recognizing our disadvantages, let's get to know the other party of this negotiation, China H group.

Fig. 4 – 7 shows the other party's advantages analysis.

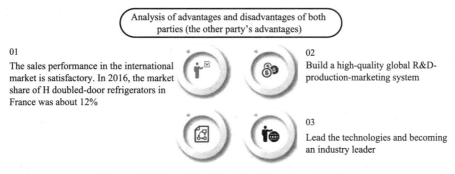

**Fig. 4 – 7　The Other Party's Advantages Analysis**

First, the information we obtained shows that the sales performance of the other party in the international market is very satisfactory. In 2016, the market share of the other party's H doubled-door refrigerators in France was about 12%. On the premise that other competitive brands work together to develop the French market, this data is still relatively competitive.

Second, China H group has built a high-quality global R & D, production and marketing system, which can be said to be very integrated.

Third, it is a leader in technologies. Of course, according to the information received, we know that the other party also has disadvantages.

We can understand each other's disadvantages as one of the breakthrough points in the negotiation.

Fig. 4 – 8 shows the other party's disadvantages analysis.

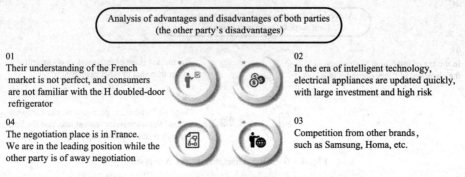

**Fig. 4 – 8  The Other Party's Disadvantages Analysis**

First, their understanding of the French market is not perfect. It can be said that consumers are completely unfamiliar with these two new products of H doubled-door refrigerator. Our investment is also a very urgent need for them to broaden channels.

Second, like our disadvantages, in the era of intelligent technology, the upgrading of electrical appliances is very fast, the investment is very large, and the risk is also very high.

Third, compared with other brands, although their market share is 12%, there is no lack of fierce competition from other brands, such as Samsung in the Republic of Korea and Homa in China.

Fourth, the main reason for their disadvantages is that the place of negotiation is in France. This is a local negotiation for us, and we can be in a leading position.

Source: Group 3, Class 1621, International Business Major, Guangxi University of Finance and Economics, Speaker: Huang Qingmiao.

# Group-based Comprehensive Simulated Negotiation

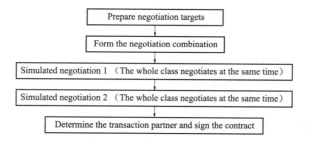

## Prepare Negotiation Targets

### 1. Schedule

It takes about 3 class hours to prepare the contents of negotiation targets. One class hour is for the group to unify, familiarize and master the contents of negotiation targets. We should strive to achieve the unity of all members on every detail, so as to avoid different or even contradictory positions among members during the negotiation.

### 2. Contents

The contents include transaction prices, trade terms, transaction quantity, payment methods, delivery dates, modes of transportation, after-sale service, etc.

Discuss and formulate the specific objectives of simulation negotiation according to the negotiation scheme, international trade contracts, market research and technical advantages. The negotiation objectives shall note that the transaction price is linked to various factors such as trade terms, transaction volume, payment methods and after-sale service, etc. If these factors change or there is room for adjustment, the different transaction prices under different circumstances shall be listed.

### 3. Adjustment Items

The objectives shall include concession items and non-concession items, such as trade terms, packaging, delivery time, payment methods, transaction quantity, quality standards, inspection methods, after-sale service, etc, so that the negotiators can clarify which aspects should be firm and which aspects can be discussed and coordinated.

### 4. Use Special Terms

1）Confirmation of delivery date. That is to say, in case of delay, the seller must compensate a certain amount of money.

2）Confirmation of quality terms. That is, the required quality is a certain parameter. If it is lower than the parameter, the buyer can return or ask for compensation. If there are clauses in the iron ore sales contract that "If the iron content exceeds 1% of the basic iron content（62.5%）, the price per ton shall be increased by _____ yuan. If the iron content is lower than 1% of the basic iron content（62.5%）, the price per ton shall be reduced by _____ yuan." Then, such terms should be included in the negotiation objectives.

3）Confirmation of the delivery standard. It regulates whether to deliver according to the quality inspection document standard or the sample standard. After confirmation, if the delivery does not meet the requirements, the buyer can return or replace the goods.

4）Document preparation and confirmation. It is stipulated that if the buyer's taking delivery of the goods is affected due to the error in the documents made by the seller, the seller shall bear all responsibilities.

5）Confirmation of packaging terms. That is packaging standards, packaging stickers and other issues. In case of inconsistency, the buyer may claim a certain amount of money.

6）In terms of after-sale service, what different requirements may be put forward by the other party? If the buyer requires the seller to send more than one technicians to the buyer's country for a long time, how should the salary and various expenses of the technicians be solved? Will each party bear a part or will one party bear it alone? If both parties bear a part, how much should they bear respectively?

If the above special terms can be agreed to be written into the contract, they may become a useful tool to get orders. However, we also need to add some restrictive conditions to protect our interests to the greatest extent. For example, in terms of delivery dates, it can be added that "In case of delay caused by force majeure such as politics and extreme weather, both parties shall negotiate the new delivery date with mutual understanding." In terms of quality terms, it can be added that "The inspection method is…And the results obtained by other inspection methods are not accepted." In terms of document preparation, it can be added that "If the buyer fails to provide information in time, or provides wrong information, or the documents made in full accordance with the buyer's intention have problems, we will not be responsible." In terms of packaging terms, we can add that "We will take photos at the time of delivery. If the delivery is intact and meets the requirements, we will not be responsible for the problems during transportation. "

Source: *JAC Foreign Trade Reference Book—JAC and His Foreign Trade Story*, edited and adapted by the author.

## Form the Negotiation Combination \\\\

Firstly, the combination of a buyer's group and a seller's group is a negotiation combination.

Secondly, after each group completes negotiations with two negotiating parties respectively, it will determine which party to sign the contract with. Both the buyer and the seller are in the position of a two-way choice. The final result is that some groups may be able to get two orders, while others can not get a single. Such a result will test not only the strengths, but also the degree of preparation, the degree of fit between the two parties, the actual play and other factors. This is very similar to the actual international business negotiation. Therefore, students need to be fully prepared to get at least one order.

## Comprehensive Simulated Negotiation \\\\

### 1. Time: 8 Class Hours

Among them, each group will complete negotiations with two negotiating partners successively, and the negotiation time with each negotiating partner is 4 class hours. The whole class will negotiate at the same time and complete the first negotiation within 4 class hours. After that, every group exchanges negotiation partners and conduct a new negotiation for another 4 hours.

### 2. Simulation Negotiation Scene Setting

The place of negotiation is set as the place of foreign buyer/seller, that is, China is the away negotiation. The scene of simulated negotiation is set as follows.

1) Scene of negotiation starting. The simulated negotiation starts when both parties enter the conference room and shake hands at the negotiation table.

2) Negotiation process is as the following.

● Host. The host should have a chairperson to welcome guests, introduce company members, briefly explain negotiation objectives and express good wishes. Pay attention to introducing yourself (position, name) before making a welcome speech.

● Guest's speech. One representative of the guest makes a speech to thank, introduce company members, explain negotiation objectives and express good wishes. Pay attention to introducing yourself (position, name) before saying thank you.

● Formal negotiation. After the above sessions are completed, the opening statement, quotation and counter-offer sessions will be arranged. The product introduction can be freely selected in a certain session. It does not specify which party will quote first, but each team will decide whether to quote first according to their own strategies.

● Record. During the negotiation, each group should have a member to record the negotiation contents. The focus of the record is not what everyone says, but the topics and clues of the negotiation, for example, what is the first topic, which party proposes, how the other party responds, which party wins, whether there is still room for research, whether it can be used as a chip in the next round of negotiation, and so on. The record is very important, which is the key for the negotiation teams to clarify their ideas and formulate the next round of negotiation plan after the fierce debate. Therefore, the recorder needs keen insight, rigorous judgment and agile recording ability.

● Adjournment. After the negotiation has made phased progress or reached an impasse, the meeting shall be adjourned to readjust ideas, find information, summarize and formulate the next plan. During the negotiation process, it is not necessary to stop or continue the negotiation according to the class time. The adjournment, restart and other sessions of the negotiation shall be determined according to the contents and process of the negotiation. During the recess, you can attend the negotiations of other groups.

● Others. In the process of simulated negotiation, a large number of auxiliary materials need to be used and records need to be made, so there is no need to be out of manuscript in the process of a simulated negotiation.

### 3. Determine the Transaction Partner and Sign the Contract

Time: 2 class hours.

After the completion of the two negotiations, each group selects the transaction partner from the two negotiation parties, and then both parties sign an international trade sales contract. After the contract is signed, give it to the teacher.

For the group that has not received an order, it is necessary to write a report to explain the group that intends to sign the contract with it, and list the negotiation results of both parties one by one with reference to the form of the contract (including the contents agreed and the contents not agreed).

### Negotiation Skills

### 1. Steps of Answering

The steps of answering are "Agree—But—Express."
Please refer to the answer steps and examples in Chapter 2 personal rapid response practice.

### 2. Building of Opening Atmosphere

We have learned in the theoretical courses that the beginning of negotiation includes a natural and friendly style, an indifferent and repressive style, a high-profile style, a delayed style, a perfunctory style, etc. In most cases, people will start with a natural and friendly style. Only in special cases, such as too much pressure from the other party, or we need to show a special position, can we use other kinds of beginning.

The creation of the opening atmosphere often lies in the five-minute "chat" time before the formal negotiation. Even with a natural and friendly start, the choice of topic will affect the effect of construction.

In a class practice, the teacher asked the students to think in groups: when your clients came from different countries, what and how would you talk with each other during these five minutes? The students' answers were about weather, sports and life, as well as economy (from the perspective of economics), politics and religion. In terms of expression, some appreciate each other and some make a comparison with China. Well, here, the teacher's comments are as follows.

In terms of contents, it's good to talk about weather and sports, because it's relatively simple, easy to resonate and easy to talk. It is too difficult to talk about the development of the two countries from the perspective of economics. After all, everyone is a businessman, not an economist. With such a difficult topic, it's possible to see an awful silence. Politics and religion are topics that should be avoided. There are many taboos and differences are easy to arise.

In terms of expression, it is very good to mention each other's advantages in the way of appreciation. Everyone is very happy. It is easy to form resonance, build trust or appreciate each other. For example, some students talk about German architectural style with German merchants, talk about *How Steel is Refined* with Russian merchants, and talk about red wines with French merchants. Merchants will be very interested. However, if a group who is talking with British businessmen says, "The weather is very good and sunny these days. It is said that London is a foggy city. Where do you think the weather is better?" What's the problem with this statement? First, the purpose of our chat is to establish resonance, and comparison can easily lead to differences. Second, you beat the other party by comparison. Will they be happy? The other party feels bad before the negotiations start, which is particularly unfavorable to the follow-up negotiations.

---

**What's Wrong With the Following Statement**

"What do you think 'one belt, one road' has brought to your company?"

What's wrong with this statement?

First, in terms of contents, it is too macro and abstract, and in this way the other party needs to think and summarize. The possibility of no-answer silence is very high, and the topic is difficult to continue easily.

Second, this statement has a sense of superiority and is easy to bring unhappiness to the other party. China is developing rapidly. When we communicate with foreign clients, it is easy to inadvertently show a sense of superiority. If such an atmosphere is not deliberately created to put pressure on the other party, the contents of a sense of superiority should be avoided as far as possible, so that negotiations and cooperation can be carried out in an equal atmosphere.

---

## 3. Offering Strategy

### 3.1   Consider Whether to Quote First or Not

Although the seller usually quotes in advance, this is not a rule. Quoting first and quoting second have their own advantages and disadvantages respectively. The party who offers first can set a maximum or minimum price for the negotiation, and can also show their own styles and attitudes. However, if the first bidder is a green hand and doesn't know much about the market, she/he may also quote a wrong price which is too low or too high, either expose her/his ignorance or make the other party feel insincere. The late bidder can figure out the other party's position and attitude and make use of them later, but he also lost the opportunity to control the situation.

### 3.2    Psychological Factors Need to Be Considered When Quoting

#### 3.2.1    Division quotation method.

If the total price is large, you can consider of splitting the total price by unit, and the smaller u-nit price will be more acceptable. For example, a high-end cold and warm air conditioner costs $899. If we explain, "The service life of the air conditioner is 12 years, that is, 4,380 days, the daily cost is only $0.2, that is, only one tenth of the price of French fries, you can enjoy the cool in summer and the warm in winter." If the other party says, "although your air conditioner is good, it is $80 more expensive than others," then we can say that it is only $0.01 more per day, less than a potato chip, but we can enjoy better quality and more functions, which is completely worth it.

#### 3.2.2    Mantissa quotation method.

This is a common quotation method in life, that is, through the setting of mantissa, the price presents a feeling of cheaper or more expensive. For example, for ordinary goods, if the price is 1.99 yuan, 9.9 yuan and 99 yuan, it will make people feel cheaper than 2 yuan, 10 yuan and 100 yuan respectively. For luxury goods, the psychological difference between 20,000 yuan and 19,999 yuan is also far greater than 1 yuan, which makes people feel very cost-effective, as the grade is one level higher, but the price is not much expensive.

#### 3.2.3    Additive quotation method.

Some goods can be sold in sets or individually. The price of the complete set of goods is high. If it's directly recommended, it may exceed the other party's expectations and the success rate is not high. However, if the other party is willing to accept one or two of the items in the whole set, we can recommend one or two of them, and then slowly add other items. When the other party is willing to buy nearly the whole set of items, it is likely for them to buy the whole set together.

#### 3.2.4    Positive and negative prices.

Each kind of goods or services has many features and parameters. Then, which point should we choose to introduce in order to arouse each other's resonance or interest? This needs preparation and research, and may need some luck. For example, chemical products have purity, moisture content, impurity content, heavy metal content, free formaldehyde content and other indicators. Different clients pay attention to different indicators, which has a great relationship with how they use the products. It is a positive price when we present the characteristics to the other party, which she/he cares about and interested.

#### 3.2.5    Integer strategy.

At the end of the negotiation, i.e., the transaction session, the other party is still hesitating. You are the seller, and the unit price is $269/box. The planned order quantity of the other party is 60 boxes, that is, the total price is $16,140. If you take the initiative to use the integer strategy, that is, the transaction amount is $16,000. Can you speed up the process of transaction? The probability of success is high.

## 4. Have Certain Professional Qualities and Abilities

When choosing partners, we will consider many factors, including whether the other party and the other party's company are reliable, the other party's strengths and professionalism, the other party's cooperation attitude, the other party's concerns about future development, etc.

So, what are the aspects of professional ability? This requires an in-depth understanding of the upstream, middle and downstream structure of the product industrial chain, including raw materials, manufacturing process, management and trade, as well as technical indicators, performance, appearance, advantages, application, certification, innovation, import and export characteristics and cases, international market dynamics and other details. Taking textile and garment as an example, professionals should understand the following knowledges.

Firstly, it's about the industrial chain structures and contents of the textile and garment industry, as shown in Table 5 – 1.

**Table 5 – 1　Industrial Chain of Textile and Garment Industry**

| Section | Main Category | Detail | Influence Factor |
|---|---|---|---|
| Upstream | Natural fiber | Cotton | Equipment automation Capability of enterprise R&D ability Cost and scale Stability of order |
| | | Hemp | |
| | | Silk | |
| | | Woolen | |
| | Chemical fiber | Regenerated fiber | |
| | | Synthetic fiber | |
| Midstream | Textile process | Yarn is processed into embryo cloth | |
| | | Fabric printed or printed from embryo cloth | |
| | Garment finished product production | — | |
| Downstream | Garment management and trade | — | Brand Design of product Channel Supply chain |

Secondly, it's about the main technical indicators of fabrics and clothing. The main technical indexes of textile fabrics include length index, warp and weft density, width and gram weight. The main technical indexes of clothing include fabric shrinkage, fabric sewing strength, fabric breaking strength, fabric tearing strength, fabric color difference, etc.

Thirdly, it's about the classification and grade of clothing. According to GB 18401 – 2010 National General Safety Technical Code for Textile Products, clothing has class A, class B and class C. Class A refers to products for infants that under 36 months old. Class B refers to products in direct contact with the skin. Class C refers products that does not directly contact the skin. In addition, garments are divided into first-class, second-class and third-class according to appearance, specification, color difference, defects, sewing and other indicators, which is convenient for both parties to use when signing trade contracts. According to the length of burning time, all textile fabrics are divided into grade 1, grade 2 and grade 3 in the Federal Regulations Code of the United States, that is, normal flammability, medium flammability and rapid flammability.

Students who choose a clothing theme need to have an in-depth understanding of the products involved according to the above basic contents, so they can have certain professional qualities and abilities. Students who choose other topics can learn from the above framework and ideas to understand the products.

---

**Not Good at English, but Performance Ranked First**

There is a young lady in our company who is not good at English and even barely has basic communication ability. However, her performance is No. 1 in our company.

How can she do that?

Source material!

She has nearly 20GB of product pictures and videos, while others are less than 3GB, which was given by the company when they joined the company.

The young lady bought a digital camera with her first month's salary. She took pictures every time she delivered goods. She asked the workers which part of the equipment was the most successful, which part was reworked, why it was reworked and which part was unique. She took pictures one by one, and then went to sort and record it.

When she was chatting with clients, there were always pictures and videos. She knew all kinds of product features and used cases like the back of her hand. Do you have any?

Source: *JAC Foreign Trade Reference Book—JAC and His Foreign Trade Story*, edited and adapted by the author.

---

## 5. Horizontal Negotiation and Vertical Negotiation

Horizontal negotiation means that as the order of main issues of the negotiation are determined, we start to discuss the predetermined issues one by one. When there are contradictions or differences on a certain issue, we will put this issue behind and discuss other issues firstly.

The discussion goes on again and again until all the contents are settled. For example, in the negotiation of capital lending, the negotiation contents should involve issues such as loan currencies, amount, interest rates, loan terms, guarantees, repayment and the grace period. If the two parties can not reach an agreement on the loan terms, this issue can be put behind, and other issues like guarantees and repayment will be discussed. When the other problems are solved, we can go back to the problem of loan terms. The core of this negotiation method is flexibility. As long as it is conducive to the settlement of the problems, the terms discussed can be adjusted at any time with the consent of both parties. This method can also be adopted: put forward the relevant key points together, discuss and study them together, so that there is room for negotiation and compromise between the problems you talk about, which is very conducive to the solution of the problems.

For example, when the other party says, "We can talk to you, but the problem is that we want to hold an annual sales conference in New Orleans. If you want to become our supplier, you must hand in the samples before the 1st day of the month on which the sales conference is held, otherwise, we don't need to waste time. " You can not accept this request and you are not sure if it's the other party's real bottom line. Can the negotiation continue? You can say, "I know this is important to you, but we might as well put this issue aside and discuss the details of the work. For example, do you want us to use union employees? What do you suggest about payment?"

Vertical negotiation refers to discussing each issue and clause one by one after determining the order of main issues of the negotiation. When an item is discussed, never put it behind even there are some problems until the end of negotiation. For example, in a product transaction negotiation, after the two parties determine several main contents such as the price, quality, transportation, insurance and claims, they begin to negotiate the price. If the price can not be determined, no other terms will be discussed. Only after the price is settled, can the two parties discuss other issues.

Horizontal negotiation and vertical negotiation have their own advantages and disadvantages. For example, horizontal negotiation is a more flexible way, which is conducive to both parties' understanding of various issues, but it may also increase the degree of bargaining between both parties, and even make the topic gradually deviate from the main line and pay too much attention to some non-key issues. Vertical negotiations can simplify complex issues and avoid the disadvantages of multi-head containment and negotiation without resolution. However, it will also lead to too rigid agenda setting, lack of mutual accommodation in discussing issues, and failure to give full play to the imagination and creativity of negotiators. It may also seriously affect the negotiation process because of an impasse.

## 6. Actions Speak Louder than Words

When we bargain with each other, our statements should be based on facts. For example, in the negotiation of technology introduction, if we want to ask the other party to reduce the price, we should be able to list the reasons for reducing the price. For example, some of these technologies are newly developed and some are about to expire. How much is the other party's annual investment in R & D, how many achievements, and how much is the average of each technology? What is the reference price among peers?

## 7. Judge Each Other's Concerns and Decisions Through Micro Expression, Body Language and Other Information

In the process of communication, it is very important to find a breakthrough point, especially when you find that the other party has not entered the state of negotiation. To cooperate, there must be resonance. So, how can we find the topic that the other party pays attention to? The obvious change of the other party's expression or action is an important signal.

## 8. Topics or Statements That Should Be Avoided

8. 1   Directly indicate that these ares the company's regulations and industry rules.

For example, in terms of payment method, if you directly tell the client: "Our company always only accepts irrevocable L/C payment at sight. " What's wrong with this? The statement itself is not wrong, but it might make the other party unhappy. If you are the other party, do you think it is necessary to continue the negotiation? They might answer: "Our company always only accepts T/T at sight," and then stop the talking. In a word, both parties do business in order to achieve cooperation and meet their needs, rather than "I'm the boss, you yield to me" unless one party has no choice. In fact, when clients have different requirements for payment methods, we can consider the specific needs of the other party.

1) If clients prefer to save costs, T/T is a better way.

2) If clients are more concerned about the liquidity of funds, they prefer the payment method of letter of credit.

3) If clients need faster delivery time, letter of credit is undoubtedly not the best choice.

And, we should also clarify some special national policies and abide by the fixed payment methods formed in some industries. For example, Bangladesh requires that all imports should be L/C + CNF, and most products should be subject to a third-party inspection, SGS or BV. In the United States, the goods can be picked up by copies of straight bill of lading instead of the original ones. If you do 30% T/T in advance and 70% against copy of B/L with American clients, there will be a great risk. Turkey, India and other countries stipulate that there should be a written certificate from the original buyer for transshipment or return. If the goods detained in the port exceed a certain date, they will be auctioned, and the original buyer has the priority. In terms of industry rules, taking machinery and pesticides as examples, the payment method of machinery products is generally 30% prepayment, 70% inspection and payment at the factory, and then leave the factory. For some pesticides, 90% are put on account, dozens of days of letters of credit or even D/A or D/P.

8.2  Talk in our own language with foreign clients.

The reason is as the same as above. This behavior itself is not wrong, but the other party will feel disrespected, and it is likely to veto all our previous efforts because of this temporary unhappiness. We should understand that there is a lot of choice for foreign trade negotiatiors, and every negotiator is constantly selecting and being selected. Before signing the contract, our clients are comparing all aspects of our products, service, strengths, cooperation prospects and even reception with our competitors.

8.3  Always want to "win" the clients in the bargaining.

When the clients put forward a request that is impossible to achieve, how would you respond? Would you smile and understand them with a peaceful attitude, or entangle in this detail and tell the other party that it is impossible. They can't get this condition anywhere. In this way, you feel comfortable while the other party doesn't.

8.4  Talk too much and disclose something that is not supposed to say.

Something should be said is said, and something not supposed to be said is said as well. In this situation, the questions that the other party haven't thought of are proposed. Of course, this approach has two sides, and clients may think that we are more honest and reliable partners.

However, when the negotiation is basically completed and the signing intention is basically determined, if we inadvertently exposes the shortcomings of the product or service (especially the negligible shortcomings), the clients are likely to think, "Oh, I didn't see this problem before. Let me think about it again." Then, they might have a new choice instead of signing the contract with us.

## 9. Arrangement of the Seats

Horizontal table seats arrangement is shown in Fig. 5 – 1.

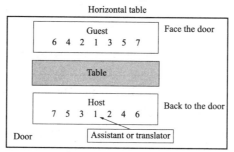

**Fig. 5 – 1  Horizontal Table Seats Arrangement**

Firstly, in the seating arrangement of the host and guests, pay attention to letting the guests sit on the side facing the door, so as to bring a sense of security to the guests. This is the host's expression of respect and sincerity to the guest.

Secondly, in the seating arrangement within the group, the main speaker should sit in the middle, the second important person should sit on the right of the main speaker, the third important person should sit on the left of the main speaker, and so on. If the main speaker has an assistant or interpreter, it can be arranged at the right rear of the main talk.

Vertical table seats arrangement is shown in Fig. 5 – 2.

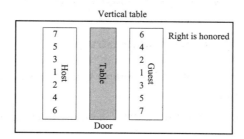

**Fig. 5 – 2  Vertical Table Seats Arrangement**

In this type of negotiation table, the principle of "right is honor" should be followed, and the guests should sit on the right side of the negotiation table to show their respect for the guests.

The seating arrangement within the group is the same as Fig. 5 – 1.

Round table seats arrangement is shown in Fig. 5 – 3.

The seating arrangement of round table negotiation is as the same as that of "long table 1." Attention should be paid to let the guest sit on the side facing the door.

Scattered seats arrangement: Both parties sit on two sides respectively, which is convenient for communication and unity within the group, but it will also produce an atmosphere of confrontation.

The scattered seats can greatly alleviate this confrontation.

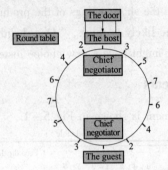

**Fig. 5 – 3　Round Table Seats Arrangement**

***

**Is It Ok Even If Sitting in the Wrong Position**?

Neil Renberg, a famous American negotiation expert, was once invited to participate in the negotiation between labor unions and managers, as the negotiator of the manager. After being introduced, the union representative asked Neil Renberg to sit opposite them. But Neil Renberg sat with union representatives. Union representatives looked at Neil Renberg with a strange expression and hinted that he was in the wrong seat, but Neil Renberg ignored it.

Shortly after the negotiation began, the trade union representatives almost forgot that Neil Renberg was on behalf of the manager. They listened carefully to his analysis, opinions and suggestions, just like listening to their own opinions and suggestions. The atmosphere of confrontation decreased greatly, while the atmosphere of harmony increased considerably. Nierenberg's choice of negotiating seats can be said to have contributed to the complete success of this negotiation.

***

## Possible Difficulties and Solutions in Negotiation

### 1. The Other Party Uses the Shadow Strategy

The other party tells us that one of our competitors (company B) provides more favorable conditions. There will be three situations.

1) We know company B, and company B has stronger strengths and better reputation than our company.

2) We know company B, but the strengths of company B are similar to that of our company, or worse.

3) We haven't heard of company B.

So, in these cases, what should we do?

Firstly, when making a statement, we can use the three steps of "Agree—But—Express" to make the other party feel psychologically being identified first, and then they would listen to our statement. Secondly, we need to express our views according to the 3 situations above.

In the first situation, congratulate the other party on getting the offer of a large company. Then we can tell them a large company has the characteristics and more rules than small companies. We can ask the other party about the specific trading conditions of company B, and then list the reasons of its disadvantages. Such as rigid behaviors, high order quantity, inflexible payment methods, inappropriate delivery time and place, and even it can be said that large companies can clearly put forward higher requirements. Why do they reduce their conditions? Is there any scandal recently, and so on. Finally, let's explain our advantages, like "You see, we have so many resonance in business and strategy. " "Our service is supreme, and our raw materials and technology are at the same level as that of company B, etc. " It should be noted that the other party's proposal of company B may also be a shadow strategy. It may not be true. Don't mess up when we hear that a large company is competing with us.

In the second situation, company B has the same strengths as ours, or worse. Our attitude is not nervous, but we should also pay attention to it and stick to our position. First of all, the conditions from company B are really good. However, there will be shortcomings. Company B has the same strengths as ours, or worse. What does the other party like? Have you talked to them specifically? Do you learn their situation on the spot? What is their business philosophy? Where are the key markets? What are trade habits? …Let the other party hesitate about the cooperation, and finally show our sincerity and list our advantages.

In the third situation, we haven't heard of company B. Then, we can ask the other party about the specific situation of company B. Which country is the company in? How many years has it been in this industry? What is the main business? What are the business characteristics? Is there a fixed partner? …In contrast, our cooperation is more secure and trustworthy. Of course, we should not make concessions easily.

## 2. Competitor's Products Have More Functions with Higher Price

The other party suggested that the products of other companies have one more function than ours. At the same time, the price is also slightly more expensive. For example, the product is a food processing machine. The other party says, "We are not very satisfied as your product does not have the 'inching cleaning' function. This sample from other companies is slightly more expensive, but it has this function. "

What should we do in this situation? First of all, we need to clarify our own position. Think about two questions. (1) Is it like the saying goes that "cheap things are not with good-quality, and good-quality goods are not cheap?" Our price is a little lower. Is the quality worse? (2) Is it still possible for us to get the deal? Are clients really attracted to that product?

The first question: every penny counts. Our products are a little cheaper, and the design and qua-lity will be a little worse. The second question: Other people's products are better, which does not mean that clients must sign the contract with other people. Otherwise, how could they come here to negotiate with us? Perhaps the client is comparing the two parties, or just trying to lower the price. In a mature and diversified market, it is normal for products to have different prices and quality. They also correspond to different client groups. The sales volume of higher-end products is not necessarily larger, and the market potential of products with appropriate prices is not necessarily smaller. Therefore, there is a possibility of transaction.

So what should we do next to facilitate the transaction?

Firstly, we should restore our confidence in our products, including our own confidence and the other party's confidence. You can ask the other party, "What do you think is the core value of this kind of products?" After all, the price difference between the two products is not big, so the extra function will certainly not be the core function. In the negotiation of food processer, the core value of the product is the motor, cutter head, stainless steel and glass, and the "inching cleaning" function is not the core component.

Secondly, analyze whether the so called non-existing function can be realized by other existing functions? Is this function necessary? Do consumers care about this function very much? …. For example, the "inching cleaning" function of the food processing machine can be realized through the "fruit and vegetable juice" function. Besides, "inching cleaning" is not a function that consumers care about very much.

Finally, reconfirm that our products are value for money. This requires analysis of products and markets, such as cost performance, other characteristics, certificates, raw material advantages, target markets, past sales, etc. In short, after comparison, we can turn the other party's attention back to our products.

Being compared are very common in negotiations, and the examples given by the other party are likely to be better than ours. At this time, the most important thing is to have confidence in yourself and negotiations. "Not confused—analyze products—analyze markets—regain confidence—facilitate transactions is a very effective step.

### 3. The Negotiation Is Basically Completed, But the Other Party Have Concerns

If this is the first cooperation, and the other party is worried about whether you can keep our promise and perform normally. The terms you quoted are reasonable and meet the requirements of transaction, but the other party is always hesitant.

In regard to concerns, we should understand that it is normal to have concerns. After all, international business risks are great. But should we say that it is difficult to eliminate concerns and promote cooperation? The answer is no. It is normal for international business to cooperate in concerns. No matter who you cooperate with, you have concerns. What we have to compete with our competitors is not only the hard power such as the company's assets and reputation, product quality, service and cooperation conditions, but also the soft power such as reception, communication and cooperation blueprint. The elimination of concerns is a basic content of communication. Trust is the foundation of cooperation. Eliminating concerns and building trust is also a basic task of the negotiations.

So, how to eliminate concerns? Firstly, make clear what are the other party's concerns? Secondly, make a reasonable explanation. Recommended ideas are as below.

1) I understand your concerns. If I were you, I would have these concerns.

2）What we care about is…It's not good for me to break my promise.

3）Cooperation is win-win for us.

4）List facts and data (if any).

## 4. The Other Party Is More Prepared

The other party is more prepared than us or has stronger momentum. In short, the whole situation is basically controlled by the other party. In the actual negotiation, the party who loses initiative in the negotiation often suffers losses, with the reason that he doesn't clear his mind in time and signed the contract with the conditions favorable to the other party.

However, this situation is really common, as the situation of equal strengths do not often occur . What should we do if we are in a weak position in the negotiation?

Firstly, never let the other party lead us by the nose. We should clear our minds in time and readjust our thinking with the opportunity of adjourning the meeting.

Secondly, we should jump out of thinking limitation with the help of reverse thinking. For example, we have to think, what they say sounds reasonable. Is it really reasonable? They say it won't work. Is it true? They say others have taken the lead. So don't we have a chance? They talk about the advantages of competitors, and say they are really so powerful. Do we have no room for maneuver? … Ask questions and return to the negotiating table only after you have an answer.

---

**Communication and Speaking Skills**

When you (the supplier) asked the other party, "where did you purchase before?" The other party replied, "I purchased in ×× country, it's the first time in China." What would you say next?

First, you can take over the other party's topic and said that I was glad that the other party could come to China, introduce China's advantages in manufacturing and industrial chain, and confirm that the other party can find satisfactory partners in China.

Second, consult the other party about some details of previous procurement. You can tentatively ask what aspects of previous cooperation you are not satisfied with, or make some comparison according to the other party's description.

Note: During the negotiation, the communication between both parties should achieve three purposes: obtaining information, expressing information and building trust. Therefore, in communication, we should not only express our position and show the prepared materials, but also be good at asking, so as to obtain more information and improve the efficiency of communication.

---

## Common Problems in the Simulated Negotiation

### 1. The Answer Hasn't Been Considered and Discussed

The response is made too fast without thinking, query of facts and data and tacit understanding of the team, so that many meaningful topics stop after only one round of debate. It is a pity that there is no following, no further research and utilization.

Example 1:

Buyer says: "Do you have any other customers in France?"

Seller says: "Our company has entered five major sales channels in France and occupies 12% of the French market. "

The students in the group of buyer don't know what to answer and the communication stops.

Has the seller's answer been impeccable and completely gained the upper hand? In fact, there are many clues to be discussed. For example, is the 12% market share a rising trend or a declining trend? What is the strength and market share of the seller's competitors? The seller has a lot of market share. Have there been any negative reports? …If the buyer is not fully prepared, they can write down the topic and conduct further research during the adjournment. It can be used as a new topic/chip in the next negotiation.

Example 2:

When the other party makes a request about 20 pieces of package free of cost (unexpected request), a negotiator says " No" immediately. This is not correct as the answer of the unexpected question should be discussed by the group.

Correct method: When an unexpected problem occurs, the team should stop talking and have an inside-team discussion about the feasibility or asking for something else as corresponding exchange conditions. If an oral discuss is not necessary, the teammates should give a sign to each other before answering.

Example 3:

When the other party claimed that our share price fell and questioned our operation conditions, a negotiator replied immediately that our operation went well.

Correct method: Firstly, we need to check the real-time share price and market prices of ourselves. Secondly, check the real-time share price and market prices of the other party. Thirdly, review the lately important events of both parties in the materials we brought in the meeting room.

## 2. There Is No Step or Focus in the Speech

The train of thought is scattered. Correct method: List the outline or key words on the paper and make clear what to say firstly, secondly and thirdly in the mind.

Example: The other party pressures us with the bribery event in 2008 and question our credit. We can reply in this way.

Firstly, to be sorry on the event. Secondly, indicate that it is an occasional event which will never change our company's principle of fair and honest. Thirdly, hope that Chinese side can treat it objectively as Australian side had performed excellently and the cooperation of both sides was good for many years.

## 3. The Speech Is Not Targeted

Selectively avoid seeing and hearing, the speech is not targeted, and the answer does not correspond to the topic thrown by the other party.

Example:

Buyer says: "Your price is too high. We can't sell it at this price. We'll lose money."

Seller says: "Our raw materials are from domestic A-level production areas. Our products have been certified by China and Europe and have absolute advantages in quality."

What are the questions in the seller's answer? The contents of this answer do not respond to the contents of the buyer, and does not focus on the subject of the other party's concern—price. In fact, the seller's answer is like a template. It has nothing to do with what the other party says, only considering what he wants to express. If you don't "hear" the other party's speech and can't answer questions and doubts, it won't work. In the negotiations, a firm position is important, but active thinking is also the key to communication, understanding, mutual benefit and win-win results, and deepening and expanding cooperation.

## 4. There Is not a Leading Person in the Team

The negotiation is in chaos as everybody answers a question together or a person's speech is often interrupted. In this case, the themes are gradually strayed or never go profound.

Correct method: assign a leading person in the team; everyone who wants to speak should give a sign in advance; the ongoing theme can't be changed or interrupted until a periodical result is achieved. The problem can be corrected when students gain more experience or the teacher makes a rule in advance. If necessary, the chief negotiator can be changed.

## 5. Underprepared

During the negotiation, one party says yes or no only, without data or facts to support. The negotiator of one party doesn't give in without any reasons, making the other party feel that the negotiators are not professional and unable to discuss further.

## 6. Incorrect Negotiation Attitude

The negotiation is too hard as some teams' standpoints are too hard on a certain issue. A principle needs to be set to prevent the problems: both parties should abide by the principle of cooperation. Giving appropriate pressure to the other party is allowed but never out of line. It is not allowed to say no all the time. If team A's reasons for the requirement are irrefutable, then team B should agree with it or call for a break.

Some teams are too easy talking and there is nothing to bargain. To prevent this from happening, both parties need to make more in-depth research on the terms of the contract, industry practices, international market and so on. After all, in international trade, problems such as concept conflict, market information imbalance, different business practices, company and individual needs, personality conflict and so on are very common.

### Communication and Speaking Skills

When the other party says, "Your sample is very good, but the size is a little too large." How do you answer it?

First of all, thank them for their affirmation of our samples.

Second, ask the other party what size they think is appropriate? What is the purpose of the goods? Then, recommend samples of appropriate size or discuss the size adjustment scheme with the other party.

Note: Many students will directly explain: "Our quality is very good." "Our user evaluation is very good." These answers are not the best, because we don't know what the customer's real needs are. The expression may not be on the point, not the content of the other party's attention, and even lead to the situation of "each says his her own things", and the communication efficiency is very low.

# Appendix 1   International Sales Contract Sample

## International Sales Contract

Contract No. :              Contract date:

Buyer:

Address:

Telephone:              Fax:

Seller:

Address:

Telephone:              Fax:

The seller and the buyer agree to conclude this contract subject to the terms and conditions stated below:

| 1. Name of Commodity and Specification | 2. Quantity | 3. Unit Price | 4. Total Amount |
|---|---|---|---|
|  |  |  |  |
|  |  |  |  |
| Total | | | |

5. Total amount in words:

6. Quantity: _____ % more or less is allowed.

7. Trade term:

☐FOB      ☐CFR      ☐CIF      ☐DDU

8. Package:

9. Shipping mark:

10. Shipment from _____ ( port of loading) to _____ ( port of destination).

11. Transshipment: ☐allowed; ☐not allowed

12. Partial shipment: □allowed; □not allowed

13. Time of delivery:

14. Insurance: to be covered by _____ for _____% of the invoice value against _____ and _____.

15. Terms of payment:

15. 1    The buyer shall pay 100% of the sales proceeds through sight draft by T/T remittance to the seller not later than _____.

15. 2    The buyer shall issue an irrevocable L/C at sight through _____ in favor of the seller prior to _____ indicating L/C shall be valid in China through negotiation within _____ days after the shipment effected, the L/C must mention the contract number.

15. 3    Documents against payment (D/P): The buyer shall duly make the payment against documentary draft made out to the buyer at sight by the seller.

15. 4    Documents against acceptance (D/A): The buyer shall duly accept the documentary draft made out to the buyer at sight by the seller.

16. Documents required: The seller shall present the following documents required for negotiation/collection to the banks.

16. 1    The full set of clean on board bills of lading.

16. 2    Signed commercial invoice in _____ copies.

16. 3    Packing list/weight [①]memo in _____ copies.

16. 4    Certificate of quantity and quality in _____ copies issued by _____.

16. 5    Insurance policy in _____ copies.

16. 6    Certificate of origin in _____ copies issued by _____.

17. Shipping terms: The seller shall immediately, upon the completion of the loading of the goods, advise the buyer of the contract number, names of commodity, loaded quantity, invoice values, gross weight, names of vessel and shipment date by tlx/fax.

18. Inspection and claims:

18. 1    The seller shall have the qualities, specifications, quantities of the goods carefully inspected by the inspection authority, which shall issue inspection certificate before shipment.

18. 2    The buyer has right to have the goods inspected by the local commodity inspection authority after the arrival of the goods at the port of destination. If the goods are found damaged /short/ their specifications and quantities not in compliance with that specified in the contract, the buyer shall lodge claims against the seller based on the inspection certificate issued by the commodity inspection authority within _____ days after the goods arrival at the destination.

18. 3    The claims, if any regarding to the quality of the goods shall be lodged within _____ days after arrival of the goods at the destination, if any regarding to the quantities of the goods, shall be lodged within _____ days after arrival of the goods at the destination . The seller shall not

---

① In order to reflet the consistence of international trade, this book follows the appellation of industry custom. The weight, net weight and gross weight in the book actually refer to mass, net mass, and gross mass.

take any responsibility if any claims concerning the shipping goods are up to the responsibility of insurance company/transportation company /post office.

19. Force majeure: The seller shall not hold any responsibility for partial or total non-performance of this contract due to force majeure. But the seller shall advise the buyer on times of such occurrence.

20. Disputes settlement: All disputes arising out of the contract or in connection with the contract shall be submitted to the China International Economic and Trade Arbitration Commission for arbitration in accordance with its rules of arbitration in _____ . The arbitral award is final and binding upon both parties.

21. Law application: It will be governed by the law of the People's Republic of China under the circumstances that the contract is singed or the goods while the disputes arising are in the People's Republic of China or the defendant is Chinese legal person, otherwise it is governed by United Nations Convention on Contract for the International Sale of Goods.

The terms in the contract based on Incoterms Rules or International Commerical Terms 2020 (INCOTERMS 2020) of the International Chamber of Commerce.

22. Versions: This contract is made out in both Chinese and English of which version is equally effective. Conflicts between these two languages arising therefrom, if any, shall be subject to Chinese version.

23. Additional Clauses: _____ (conflicts between contract clause here above and this additional clause, if any, it is subject to this additional clause).

24. This contract is in _____copies, effective since being signed/sealed by both parties.


Buyer:                          Seller:

Signature                       Signature

Date                            Date

# Appendix 2  Iron Ore Trade Contract Sample

## Iron Ore Trade Contract

This contract is made on _____

Buyer：

Address：

Telephone：

Seller：

Address：

Telephone：

### 1. Product Supply，Quantity，Delivery

1. 1   Seller's liability and product's delivery from seller's mine.

1. 1. 1   Subject to the terms and conditions of this contract, in each contract year, the seller shall sell and deliver and the buyer shall purchase, take delivery and pay for the annual quantities of product, which shall be agreed between the parties in accordance with _____ of this contract. Quantities stipulated in accordance with _____ shall be delivered by the seller to the buyer on a free on board ("FOB") spout trimmed basis at the loading port, in accordance with the loading conditions ("Delivery"). Title to product, and all risks associated thereto, that have been delivered by the seller to the buyer shall pass from seller to buyer in accordance with this contract.

1. 1. 2   The product sold and purchased pursuant to this contract shall be mined and produced by the seller at the seller's mine.

1. 1. 3   Without prejudice to the annual quantity determination procedure set forth in this contract, both parties hereby agree and acknowledge that the product quantities to be sold and purchased pursuant to this contract shall be in a volume sufficient to allow the seller to obtain financing for the seller's mine on a limited-recourse basis, pursuant to sections below.

1. 2   Notice for failed delivery.

1. 2. 1   If at any time the seller becomes aware that it will not be able to produce and deliver the amount of products set forth in a delivery schedule agreed in accordance with this contract, the seller shall promptly notify the buyer in writing accordingly and shall, in the relevant notice, set out the reasons for the expected failure to meet the relevant order.

1. 2. 2   In the event the seller fails to deliver any amount of products in accordance with this contract, the seller shall indemnify the buyer against all costs, damages, expenses, liabilities, losses, penalties or fines, including without limitation, dead freight and/or any charges in diverting the vessel, incurred by the buyer arising from the seller's failure to deliver.

1. 2. 3   Conversely, in the event the buyer fails to nominate and purchase products in accord-

ance with any schedule agreed in accordance with this contract, the buyer shall indemnify the seller against all costs, damages, expenses, liabilities, losses, penalties or fines, including without limitation, loss of sale, incurred by the buyer arising from the seller's failure to purchase.

### 2. Irrevocable Letter of Credit

2.1   Buyer's assurance on payment liability.

2.1.1   In order to secure the punctual payment of all of its payment obligations hereunder, the buyer hereby agrees to establish, until no later than _____ days prior to commencement of each quarter of the contract year, an unrestricted, non-transferable and irrevocable letter of credit, payable at sight.

2.1.2   The seller's delivery obligations related to the relevant quarter of the contract year shall only commence after the respective LC has been properly issued.

2.1.3   Each LC shall remain at all times in full force and effect until all the buyer's payment obligations regarding the corresponding quarter of the contract year have been fully satisfied, or until such other term that might be eventually agreed upon by the seller and the buyer in writing, whichever occurs first.

2.1.4   At the last business day of each and every quarter of the contract year, the buyer shall provide the seller with satisfactory evidence that the LC remains in full force and effect.

2.1.5   In the event the acceptable bank is no longer acceptable, the seller shall notify the buyer in writing so as to permit the buyer to arrange for a newly issued LC by a premier international financial institution acceptable to the seller.

### 3. Sample and Analysis

3.1   Sampling and analysing at the loading port.

3.1.1   For each shipment of products, a representative sampling as well as physical and chemical analysing shall be affected at the loading port by the seller, at the seller's own cost and expense.

3.1.2   The buyer shall have the right to appoint, at its own cost and expense, a representative (to be approved by the seller, which approval shall not be unreasonably withheld or delayed) to follow the procedures for sampling as well as physical and chemical analysing.

3.1.3   The physical and chemical analysing effected by the seller shall be the basis for the respective certificate of analysis to be issued at the loading port, at the seller's own cost and expense. The seller shall send such certificate of analyses to the buyer by e-mail, within _____ business days of the date of departure of the relevant shipment from the loading port.

3.1.4   The seller shall keep one sealed gross sample taken at the loading port for at least _____ days after the date of departure of each respective shipment from the loading port in case an umpire analysis is required for the settlement of chemical or physical analysing.

3.1.5   Sampling procedures at the loading port shall be agreed between the seller and the buyer. Physical and chemical analysing at the loading port shall be carried out in accordance with the applicable ISO standard procedures.

3.2   Sampling and analysing at the discharging port.

3.2.1 For each shipment of products, a representative sampling as well as physical and chemical analysing shall be effected at the discharging port by the buyer, at the buyer's own cost and expense.

3.2.2 The seller shall have the right to appoint, at its own cost and expense, a representative (to be approved by the buyer, which approval shall not be unreasonably withheld or delayed) to follow the procedures for sampling as well as for physical and chemical analysing.

3.2.3 The physical and chemical analyses affected by the buyer shall be the basis for the respective certificate of inspection and analysis to be issued at the discharging port, at the buyer's own cost and expense. The buyer shall send such certificate of inspection and analyses to the seller by e-mail, within _____ business days of the date of completion of the discharge of the relevant shipment at the discharging port.

3.2.4 The buyer shall keep one spot sample taken at the discharging port for at least _____ days after the arriving date of each respective shipment in case an umpire analysis is required for the settlement of chemical or physical analysing.

3.2.5 Sampling procedures at the discharging port shall be agreed between the seller the and the buyer. Physical and chemical analyses at the discharging port shall be carried out in accordance with the applicable ISO standard procedures.

3.3 Product's final analysing result.

3.3.1 If the difference between the iron content indicated in the certificate of analysis issued at the loading port and the certificate of inspection and analyses issued at the discharging port are equal to or more than 1.0 % (one percent), the samples sealed and retained for an umpire analysis by both parties shall be forwarded to an umpire analysis, internationally recognized, to be chosen by mutual agreement between the seller and the buyer, who shall analyze such samples.

3.3.2 If no analysis is performed at the discharging port or, if performed, the certificate of inspection and analyses are not issued by the buyer within _____ business days from the completion of discharge at the discharging port, the seller's certificate of analyses at the loading port shall be final and binding by the parties.

### 4. Weight Determination

4.1 Weight determination at the loading port.

4.1.1 For each shipment of products, the seller shall determine the loaded quantity of products into the relevant vessel by draft survey at the loading port, at the seller's own cost and expense, in accordance with applicable international practices.

4.1.2 The buyer shall have the right to appoint, at the buyer's own cost and expense, a representative (to be approved by the seller, which approval shall not be unreasonably withheld or delayed) to follow the procedures for weight determination.

4.1.3 The weight determined at the loading port shall be the basis for the B/L and for the certificate of weight to be issued at the loading port, at the seller's own cost and expense. Such certificate of weight shall be sent to the buyer by e-mail, within _____ business days from the date of departure of the relevant shipment from the loading port, and shall be the basis for the issuance of the respective invoice.

4. 2    Weight determination at the discharging port.

4. 2. 1    For each shipment of products, the buyer shall determine (or cause to be determined) the unloaded quantity of products from the relevant vessel by the draft survey at the discharging port, at the buyer's own cost and expense, in accordance with applicable international practices.

4. 2. 2    The seller shall have the right to appoint, at its own cost and expense, a representative (to be approved by the buyer, which approval shall not be unreasonably withheld or delayed) to follow the procedures for weight determination.

4. 2. 3    The buyer shall send the certificate of weight issued at the discharging port to the seller, by email, within _____ business days from the date of completion of the discharge of the relevant shipment at the discharging port.

4. 3    Final weight.

4. 3. 1    Final and binding weight should be decided by the following methods.

(i) If the difference between the dry weight (as defined below) of the products at the loading port and the discharging port is less than or equal to 0. 5% (zero point five percent), the weight indicated in the certificate of weight issued at the loading port shall prevail.

(ii) If the difference between the dry weight of the products at the loading port and the discharging port is more than 0. 5% (zero point five percent) or equal to 1. 0% (one percent), the arithmetic average of the weight indicated in the certificates of weight issued at the loading port and the discharging port shall prevail.

(iii) If the difference between the dry weight of the products at the loading port and the discharging port is more than 1. 0% (one percent), the two parties shall discuss and agree the final and binding weight.

4. 3. 2    If no draft survey is performed at the discharging port or, if performed, the certificate of weight is not issued within _____ business days from the completion of discharge at the discharging port, the seller's certificate of weight at the loading port shall be considered final and binding by the two parties.

**5. Price and Price Adjustment**

5. 1    Both parties agree to the following price of the product with all sums due and payable in USD: USD _____ per dry metric ton (net), payable to the seller with adjustments according to the following items.

5. 2    Price adjustment.

The price shall be adjusted every _____ (month/week). The adjustment is based on the Metal Bulletin Iron Ore Index or mysteel. com or other major international price standards. The first price adjustment will be made on _____ (date).

5. 2. 1    Fe.

If the Fe content exceeds 1% of the basic Fe content (62. 5%), the price per ton should be increased by USD _____. If the content is less than 1% of the basic Fe content (62. 5%), the price per ton should be reduced by USD _____.

If the Fe content is lower than 60% , the goods will be rejected, and the seller has the responsibilities for compensating the buyer.

5.2.2　Aluminum oxide （$Al_2O_3$）.

The content of aluminum oxide should be within 1.2% , otherwise, there will be a deduction at the rate of USD _____ per dry metric ton for each 0.1% in excess of 1.20%.

5.2.3　Silica （$SiO_2$）.

The content of silica should be within 4.5% , otherwise, there will be a deduction at the rate of USD _____ per dry metric ton for each 0.1% in excess of 4.5 %.

5.2.4　Phosphorus （P）.

The content of phosphorus should be within 0.07% , otherwise, there will be a deduction at the rate of USD _____ per dry metric ton for each 0.01 % in excess of 0.07 %.

5.2.5　Sulfur （S）.

The content of sulphur should be within 0.07% , otherwise, there will be a deduction at the rate of USD _____ per dry metric ton for each 0.01% in excess of 0.07 %.

5.2.6　Manganese （Mn）.

The content of phosphorus should be within 0.06% , otherwise, there will be a deduction at the rate of USD _____ per dry metric ton for each 0.01% in excess of 0.06%.

5.2.7　Moisture.

If the moisture content exceeds 8% , the seller will compensate the buyer for the loss by weight. However, during the rainy season （July to December every year） 12% is acceptable.

5.2.8　Physical characteristics.

（i） When the particle size of iron ore exceeds 50 mm and exceeds 5% , the price reduction ratio per dry ton is USD _____ per ton.

（ii） When the particle size of iron ore is less than 10 mm and more than 5% , the price reduction ratio per dry ton is USD _____ per ton.

**6. Overall**

6.1　The price that the buyer shall pay to the seller for the product set forth in this section shall be calculated in USD for each shipment.

6.2　Invoice.

6.2.1　The seller shall issue an invoice in respect of the relevant shipment （"invoice"） in an amount corresponding to 100% FOBST cargo value. The invoice shall be issued based on the following items.

（i） The applicable base price for the shipment as set forth in this contract.

（ii） The weight of the relevant shipment should be pursuant to the certificate of weight issued at the loading port.

6.2.2　Upon completion of loading of each shipment of products, the seller shall send the following documents by internationally recognized express courier, to the negotiating bank and the buyer, within 21 （twenty one） days after the date of departure of the relevant shipment of products from the loading port.

(ⅰ) The full set of 3 (three) clean on board bills of lading (the "B/L") are made out to send, The bills of lading should be blank endorsed, notifying party in blank and marked "freight payable as per charter party", showing the weight in WMT.

(ⅱ) The invoice at sight based on 100% (one hundred percent) of the FOBST value and weight of the cargo at the loading port, indicating the name of the carrying vessel.

(ⅲ) The certificate of analysis issued pursuant to this contract, showing actual results of chemical and physical analysis in 2 (two) originals and 3 (three) copies.

(ⅳ) The certificate of weight issued at the loading port pursuant to this contract in 2 (two) originals and 3 (three) copies.

(ⅴ) The certificate of origin in three copies issued by _____ or other authorized local authority in 1 (one) original and 3 (three) copies.

6.2.3   Within 5 (five) business days from B/L date, the seller shall provide the shipment details to the buyer via e-mail .

6.2.4   Within 10 (ten) business days from B/L date, the seller shall provide to the buyer via e-mail the dispatch details of the original documentation and the copies of all the documentation set forth in this section.

6.2.5   The relevant invoice shall be payable upon presentation of the clean documentation listed above by way of deduction of the respective amount from the LC applicable for the respective shipment.

6.3   Overpaid and underpaid amount.

Unless otherwise agreed by the two parties, any amount overpaid or underpaid in respect of the relevant shipment determined after the adjustments on the basis of final certificate of weight and analysis at the discharging port shall be reimbursed to the buyer by the seller or to the seller by the buyer, as soon as possible but no later than _____business days after the date of receipt of documents evidencing such over-payment or under-payment, without breach of this contract.

6.4   Late payment.

The relevant party shall provide the other party with a notice in writing that it has not received payment on the due date as soon as reasonably practicable thereafter. Not withstanding the delivery of any such notice, the defaulting party shall be deemed to be in default as of the date on which such party failed to make the respective payment.

6.5   Disputing payment.

6.5.1   If any party disputes any payment under this contract, the party which disputes shall pay the undisputed amount on the due date therefore and as soon as reasonably practicable give the other party reasonable details in writing of the reason for disputing the payment.

6.5.2   Any amounts payable following resolution of such dispute shall be paid within _____business days by the defaulting party and shall include interest and penalties calculated pursuant to this contract.

6.6   Payment currency.

All payments made under this contract will be made in US dollars and in immediately available cash by wire transfer.

### 7. Transfer of property right and Risk

7. 1  Property right and risk in the product delivered under this contract shall be transferred from the seller to the buyer as the product crosses buyer's vessel's rail at the loading port, provided that all products supplied in accordance with this contract will be free from liens, charges, security interest, encumbrances and unfavourable claims of any kind, except for those created by the buyer.

### 8. Force Majeure

Neither party shall be responsible for nonperformance or delay in performance under this agreement and/or any individual contract due to acts of god, civil commotions, wars, terrorist attacks or disturbances, riots, strikes, lockouts, severe weather, fires, explosions, governmental actions or other similar causes beyond the control of one party, provided that the party so affected shall promptly give notice to the other party and shall continue to take all actions reasonably within its power herewith as fully as possible.

### 9. Default and Termination

9. 1  Termination.

9. 1. 1  The party which is not in default may, by notice in writing to the party in default, terminate this contract if any of the following events of default occurs.

(ⅰ) The other party defaults in the payment of undisputed amounts due and payable under this contract exceeding in the aggregate amount of USD _____ and such default continues unremedied after the expiry of _____ days following the date on which the non-defaulting party shall have given notice of the default to the defaulting party.

(ⅱ) Dissolution or liquidation of the other party, except for voluntary dissolution or liquidation as part of a corporate restructuring, provided that the resulting entity has assumed any and all obligations of the other party.

(ⅲ) Bankruptcy or judicial or extra-judicial reorganization of the other party pursuant to applicable bankruptcy law.

(ⅳ) The other party commits any material breach of any of its obligations under this contract (other than those specifically described in item (ⅰ) of this section), including failure by the seller to supply the annual quantities in a given contract year and failure by the buyer to purchase the annual quantities in a given contract year, except if such breach is remedied by the defaulting party within _____ days from the receipt of the dispute notice sent by the non-defaulting party giving details of the breach and requiring the cure of such breach.

9. 2  Neither party may terminate this contract without a reason.

### 10. Dispute Settlement

10. 1  General.

10. 1. 1  All disputes between the parties related to any matter or disagreement arising from this contract ( "dispute" ) shall be referred by the aggrieved party to the other party by giving the dispute notice in accordance with this contract terms ( "dispute notice" ). The two parties shall, within the period of 30 (thirty) calendar days as from the receipt of the dispute notice, endeavor to come, in

good faith, to an understanding on such matter or disagreement.

10. 1. 2    If a dispute is not resolved according to the contract terms, each party shall ensure that senior executive officers of the parties meet with a view to resolve the dispute. If the dispute is not resolved within 45 (forty-five) days as from the receipt of the dispute notice, either party may require that the dispute is resolved in accordance with this contract terms.

10. 1. 3    Except where clearly prevented by the nature of the dispute, the parties shall continue performing their respective obligations under this contract until the dispute is resolved in accordance with the provisions of this contract.

10. 2    Arbitration.

10. 2. 1    In the event the two parties are unable to resolve the matter or the disagreements in accordance with the process described above, all the matters that are still in disagreement shall be submitted to the decision of, and shall be finally settled by arbitration conducted in accordance with the rules of _____, then in effect, except as herein modified by the two parties or otherwise agreed to by the two parties. All arbitration proceedings shall be conducted in _____ and shall take place in _____.

### 11. Governing Law

11. 1    This contract shall be governed by and construed in accordance with the laws of _____, without giving effect to the conflicts of laws thereof. The trade terms under this contract shall be governed by and interpreted under the provisions of INCOTERMS 2020 and its supplements in force at the date of the shipment of the products here under.

11. 2    The application of the United Nations Convention on Catracts for the International Sale of Goods is expressly excluded.

### 12. Confidentiality

The two parties hereby agree that the terms of this contract are confidential and that, except for the purpose of enforcing this contract, neither party shall disclose any of the terms of this contract to a third party other than an affiliate of a party and/or client (s). Other exceptions include such disclosure is, as evidenced by opinion of counsel, required by law or regulation of a stock exchange to which a party is affiliated or prior written approval has been obtained from the other party.

### 13. Miscellaneous Provisions

13. 1    Amendments.

13. 1. 1    Except if otherwise expressly provided herein, no provision of this contract may be amended, modified, waived, discharged or terminated, unless if made in written and signed by the two parties hereto, nor may any breach of any provision of this contract be waived or discharged by any party except with the written consent of the other party.

13. 2    Entire Contract.

13. 2. 1    This contract contains the entire agreement of the two parties and supersedes any previous written or oral agreement between the parties in relation to the matters dealt with in this contract and exclude of any terms implied by law which may be excluded by this contract. The parties

acknowledge that they have not been induced to enter into this contract by any representation, warranty or undertaking not expressly incorporated into this contract.

13.3   Taxes.

13.3.1   Any and all taxes and duties levied on the products or on this contract in ＿＿＿shall be afforded by the seller.

13.3.2   Any and all taxes and duties levied on the products or on this contract out of the country of origin shall be afforded by the buyer.

Seller：                                        Buyer：

Signature                                      Signature

Date                                           Date